Springer Series in Statistics

Advisors:
P. Bickel, P. Diggle, S. Fienberg, K. Krickeberg,
I. Olkin, N. Wermuth, S. Zeger

Springer

New York
Berlin
Heidelberg
Hong Kong
London
Milan
Paris
Tokyo

Springer Series in Statistics

Andersen/Borgan/Gill/Keiding: Statistical Models Based on Counting Processes.
Atkinson/Riani: Robust Diagnostic Regression Analysis.
Berger: Statistical Decision Theory and Bayesian Analysis, 2nd edition.
Borg/Groenen: Modern Multidimensional Scaling: Theory and Applications
Brockwell/Davis: Time Series: Theory and Methods, 2nd edition.
Chan/Tong: Chaos: A Statistical Perspective.
Chen/Shao/Ibrahim: Monte Carlo Methods in Bayesian Computation.
David/Edwards: Annotated Readings in the History of Statistics.
Devroye/Lugosi: Combinatorial Methods in Density Estimation.
Efromovich: Nonparametric Curve Estimation: Methods, Theory, and Applications.
Eggermont/LaRiccia: Maximum Penalized Likelihood Estimation, Volume I:
 Density Estimation.
Fahrmeir/Tutz: Multivariate Statistical Modelling Based on Generalized Linear
 Models, 2nd edition.
Fan/Yao: Nonlinear Time Series: Nonparametric and Parametric Methods.
Farebrother: Fitting Linear Relationships: A History of the Calculus of Observations
 1750-1900.
Federer: Statistical Design and Analysis for Intercropping Experiments, Volume I:
 Two Crops.
Federer: Statistical Design and Analysis for Intercropping Experiments, Volume II:
 Three or More Crops.
Ghosh/Ramamoorthi: Bayesian Nonparametrics.
Glaz/Naus/Wallenstein: Scan Statistics.
Good: Permutation Tests: A Practical Guide to Resampling Methods for Testing
 Hypotheses, 2nd edition.
Gouriéroux: ARCH Models and Financial Applications.
Gu: Smoothing Spline ANOVA Models.
Györfi/Kohler/Krzyżak/ Walk: A Distribution-Free Theory of Nonparametric
 Regression.
Haberman: Advanced Statistics, Volume I: Description of Populations.
Hall: The Bootstrap and Edgeworth Expansion.
Härdle: Smoothing Techniques: With Implementation in S.
Harrell: Regression Modeling Strategies: With Applications to Linear Models,
 Logistic Regression, and Survival Analysis
Hart: Nonparametric Smoothing and Lack-of-Fit Tests.
Hastie/Tibshirani/Friedman: The Elements of Statistical Learning: Data Mining,
 Inference, and Prediction
Hedayat/Sloane/Stufken: Orthogonal Arrays: Theory and Applications.
Heyde: Quasi-Likelihood and its Application: A General Approach to Optimal
 Parameter Estimation.
Huet/Bouvier/Gruet/Jolivet: Statistical Tools for Nonlinear Regression: A Practical
 Guide with S-PLUS Examples.
Ibrahim/Chen/Sinha: Bayesian Survival Analysis.
Jolliffe: Principal Component Analysis.

(continued after index)

Charles F. Manski

Partial Identification of
Probability Distributions

Springer

Charles F. Manski
Department of Economics
Northwestern University
2001 Sheridan Road
Evanston, IL 60208-2600
USA
cfmanski@northwestern.edu

Library of Congress Cataloging-in-Publication Data
Manski, Charles F.
    Partial identification of probability distributions / Charles F. Manski.
        p. cm. — (Springer series in statistics)
    Includes bibliographical references and index.
    ISBN 0-387-00454-8 (alk. paper)
        1. Distribution (Probability theory)   2. Regression analysis.   I. Title.   II. Series.
    QA273.6.M294 2003
    519.2′4—dc21   2003042476

ISBN 0-387-00454-8   Printed on acid-free paper.

Printed in the United States of America.

9 8 7 6 5 4 3 2 1   SPIN 10911921

Typesetting: Pages created using the author's WordPerfect files.

www.springer-ny.com

Springer-Verlag   New York Berlin Heidelberg
A member of BertelsmannSpringer Science+Business Media GmbH

To Arthur S. Goldberger, who encouraged me to fly

Preface

Early on, when my research on partial identification was a lonely undertaking, Arthur Goldberger saw its potential and encouraged me to keep at it. He especially lifted my spirits when, reacting to a new finding that I had excitedly shown him, he remarked "now you are flying." I have not had many ways to express how important Art has been to me over the years as colleague, critic, and friend. Dedicating this book to him is one.

The enterprise has gradually become less lonely as others have become interested in, and contributed to, research on partial identification. Four of the ten chapters in this book are based on joint work with co-authors with whom I have enjoyed very productive collaborations. Chapters 3 and 4 are based on several published articles co-authored with Joel Horowitz. Chapter 5 is based on an article co-authored with Philip Cross and Chapter 9 on one co-authored with John Pepper. I have benefitted greatly from the opportunity to work with Joel, Phil, and John on these specific projects as well as from our many discussions of subjects of mutual interest.

Jeff Dominitz, Francesca Molinari, John Pepper, and Daniel Scharfstein provided detailed and constructive comments on a draft of this book, completed in the summer of 2002. I was fortunate that all four have been interested in the book and eager to help me improve its coverage and exposition. I am grateful to the students enrolled in my Ph.D. field courses in econometrics at Northwestern University in spring and fall 2002. I tried out early versions of various book chapters on the spring 2002 class and worked through the entire draft with the fall 2002 class. I am also grateful to Joerg Stoye, who read the revised manuscript.

The National Science Foundation has supported my research program through a succession of grants. My preparation of the book itself was supported in part under grant SES-0001436.

Chicago, Illinois   Charles F. Manski
January 2003

Contents

Contents

Introduction: Partial Identification and Credible Inference

Statistical inference uses sample data to draw conclusions about a population of interest. However, data alone do not suffice. Inference always requires assumptions about the population and the sampling process. Statistical theory illuminates the logic of inference by showing how data and assumptions combine to yield conclusions.

Empirical researchers should be concerned with both the logic and the credibility of their inferences. Credibility is a subjective matter, yet I take there to be wide agreement on a principle that I shall call:

The Law of Decreasing Credibility: The credibility of inference decreases with the strength of the assumptions maintained.

This principle implies that empirical researchers face a dilemma as they decide what assumptions to maintain: Stronger assumptions yield inferences that may be more powerful but less credible. Statistical theory cannot resolve the dilemma but can clarify its nature.

It is useful to distinguish combinations of data and assumptions that point-identify a population parameter of interest from ones that place the parameter within a set-valued identification region. Point identification is the fundamental necessary condition for consistent point estimation of a parameter. Strengthening an assumption that achieves point identification may increase the attainable precision of estimates of the parameter. Statistical theory has had much to say about this matter. The classical theory of local asymptotic efficiency characterizes, through the Fisher information matrix, how attainable precision increases as more is assumed known about a population distribution. Nonparametric regression analysis shows how the attainable rate of convergence of estimates increases as more is assumed about the shape of the regression. These and other achievements provide

1

important guidance to empirical researchers as they weigh the credibility and precision of alternative point estimates.

Statistical theory has had much less to say about inference on population parameters that are not point-identified (see the historical note at the end of this Introduction). It has been commonplace to think of identification as a binary event—a parameter is either identified or it is not—and to view point identification as a precondition for meaningful inference. Yet there is enormous scope for fruitful inference using data and assumptions that partially identify population parameters. This book explains why and shows how.

Origin and Organization of the Book

The book has its roots in my research on nonparametric regression analysis with missing outcome data, initiated in the late 1980s. Empirical researchers estimating regressions commonly assume that missingness is random, in the sense that the observability of an outcome is statistically independent of its value. Yet this and other point-identifying assumptions have regularly been criticized as implausible. So I set out to determine what random sampling with partial observability of outcomes reveals about mean and quantile regressions if nothing is known about the missingness process or if assumptions weak enough to be widely credible are imposed. The findings were sharp bounds whose forms vary with the regression of interest and with the maintained assumptions. These bounds can readily be estimated using standard methods of nonparametric regression analysis.

Study of regression with missing outcome data stimulated investigation of more general incomplete data problems. Some sample realizations may have unobserved outcomes, some may have unobserved covariates, and others may be entirely missing. Sometimes interval data on outcomes or covariates are available, rather than point measurements. Random sampling with incomplete observation of outcomes and covariates generically yields partial identification of regressions. The challenge is to describe and estimate the identification regions produced by incomplete-data processes when alternative assumptions are maintained.

Study of regression with missing outcome data also naturally led to examination of inference on treatment response. Analysis of treatment response must contend with the fundamental problem that counterfactual outcomes are not observable; hence my findings on partial identification of regressions with missing outcome data were directly applicable. Yet analysis of treatment response poses much more than a generic missing-data problem. One reason is that observations of realized outcomes, when combined with suitable assumptions, can provide information about counterfactual ones. Another is that practical problems of treatment choice motivate much research on treatment response and thereby determine what population parameters are of interest. So I found it productive to examine inference on treatment response as a subject in its own right.

Another subject of study has been inference on the components of finite probability mixtures. The mathematical problem of decomposition of finite mixtures arises in many substantively distinct settings, including contaminated sampling, ecological inference, and regression with missing covariate data. Findings on partial identification of mixtures have application to all of these subjects and more.

This book presents the main elements of my research on partial identification of probability distributions. Chapters 1 through 3 form a unit on prediction with missing outcome or covariate data. Chapters 4 and 5 form a unit on decomposition of finite mixtures. Chapter 6 is a stand-alone analysis of response-based sampling. Chapters 7 through 10 form a unit on the analysis of treatment response.

Whatever the particular subject under study, the presentation follows a common path. I first specify the sampling process generating the available data and ask what may be inferred about population parameters of interest in the absence of assumptions restricting the population distribution. I then ask how the (typically) set-valued identification regions for these parameters shrink if certain assumptions are imposed. There are, of course, innumerable assumptions that could be entertained. I mainly study statistical independence and monotonicity assumptions.

The approach to inference that runs throughout the book is deliberately conservative and thoroughly nonparametric. The traditional way to cope with sampling processes that partially identify population parameters has been to combine the available data with assumptions strong enough to yield point identification. Such assumptions often are not well motivated, and empirical researchers often debate their validity. Conservative nonparametric analysis enables researchers to learn from the available data without imposing untenable assumptions. It enables establishment of a domain of consensus among researchers who may hold disparate beliefs about what assumptions are appropriate. It also makes plain the limitations of the available data. When credible identification regions turn out to be uncomfortably large, researchers should face up to the fact that the available data do not support inferences as tight as they might like to achieve.

By and large, the analysis of the book rests on the most elementary probability theory. As will become evident, an enormous amount about identification can be learned from judicious application of the Law of Total Probability and Bayes Theorem. To keep the presentation simple without sacrificing rigor, I suppose throughout that conditioning events have positive probability. With appropriate attention to smoothness and support conditions, the various propositions that involve conditioning events hold more generally.

The book maintains a consistent notation and usage of terms throughout its ten chapters, with the most basic elements set forth in Chapter 1 and elaborations introduced later as required. Random variables are always in *italics* and their realizations in normal font. The main part of each chapter is written in textbook style, without references to literature. However, each chapter has complements and endnotes that place the analysis in context and elaborate in eclectic ways. The first endnote of each chapter cites the sources on which the chapter draws. These primarily are research articles that I have written, often with co-authors, over the period 1989–2002.

This book complements my earlier book *Identification Problems in the Social Sciences* (Manski, 1995), which exposits basic themes and findings on partial identification in an elementary way intended to be broadly accessible to students and researchers in the social sciences. The present book develops the subject in a rigorous, thorough manner meant to provide the foundation for further study by statisticians and econometricians. Readers who are entirely unfamiliar with partial identification may want to scan at least the introduction and first two chapters of the earlier book before beginning this one.

Identification and Statistical Inference

This book contains only occasional discussions of problems of finite-sample statistical inference. Identification and statistical inference are sufficiently distinct for it to be fruitful to study them separately. As burdensome as identification problems may be, they at least have the analytical clarity of exercises in deductive logic. Statistical inference is a more murky matter of induction from samples to populations.

The usefulness of separating the identification and statistical components of inference has long been recognized. Koopmans (1949, p. 132) put it this way in the article that introduced the term *identification* into the literature:

> In our discussion we have used the phrase "a parameter that can be determined from a sufficient number of observations." We shall now define this concept more sharply, and give it the name *identifiability* of a parameter. Instead of reasoning, as before, from "a sufficiently large number of observations" we shall base our discussion on a hypothetical knowledge of the probability distribution of the observations, as defined more fully below. It is clear that exact knowledge of this probability distribution cannot be derived from any finite number of observations. Such knowledge is the limit approachable but not attainable by extended observation. By hypothesizing nevertheless the full availability of such knowledge, we obtain a clear separation between problems of statistical

inference arising from the variability of finite samples, and problems of identification in which we explore the limits to which inference even from an infinite number of observations is suspect.

Historical Note

Partial identification of population parameters has a long but sparse history in statistical theory. Frisch (1934) developed sharp bounds on the slope parameter of a simple linear regression when the covariate is measured with mean-zero errors; fifty years later, his analysis was extended to multiple regression by Klepper and Leamer (1984). Frechét (1951) studied the conclusions about a joint probability distribution that may be drawn given knowledge of its marginals; see Ruschendorf (1981) for subsequent findings. Duncan and Davis (1953) used a numerical example to show that ecological inference is a problem of partial identification, but formal characterization of identification regions had to wait more than forty years (Horowitz and Manski, 1995; Cross and Manski, 2002). Cochran, Mosteller, and Tukey (1954) suggested conservative analysis of surveys with missing outcome data due to nonresponse by sample members, but Cochran (1977) subsequently downplayed the idea. Peterson (1976) initiated study of partial identification of the competing risk model of survival analysis; Crowder (1991) and Bedford and Meilijson (1997) have carried this work further.

Throughout this book, I begin with the identification region obtained using the empirical evidence alone and study how distributional assumptions may shrink this region. A mathematically complementary approach is to begin with some point-identifying assumption and examine how identification decays as this assumption is weakened in specified ways. Methodological research of the latter kind is variously referred to as *sensitivity*, *perturbation*, or *robustness* analysis. For example, studying the problem of missing outcome data, Rosenbaum (1995) and Scharfstein, Rotnitzky, and Robins (1999) investigate classes of departures from the point-identifying assumption that data are missing at random.

1

Missing Outcomes

1.1. Anatomy of the Problem

To begin, suppose that each member j of a population J has an outcome y_j in a space Y. The population is a probability space (J, Ω, P) and $y: J \rightarrow Y$ is a random variable with distribution $P(y)$. A sampling process draws persons at random from J. A realization of y may or may not be observable, as indicated by the realization of a binary random variable z. Thus y is observable if $z = 1$ and not observable if $z = 0$. The problem is to use the available data to learn about $P(y)$.

The structure of this familiar problem of inference with missing outcome data is displayed by the Law of Total Probability

$$P(y) = P(y|z = 1)P(z = 1) + P(y|z = 0)P(z = 0). \qquad (1.1)$$

The sampling process asymptotically reveals the distribution of observable outcomes, $P(y|z = 1)$, and the distribution of observability, $P(z)$. The sampling process is uninformative regarding the distribution of missing outcomes, $P(y|z = 0)$. Hence, the empirical evidence asymptotically reveals that $P(y)$ lies in the *identification region*

$$H[P(y)] \equiv [P(y|z = 1)P(z = 1) + \gamma P(z = 0), \, \gamma \in \Gamma_Y], \qquad (1.2)$$

where Γ_Y denotes the space of all probability measures on Y. The feasible values of $P(y)$ are the mixtures of $P(y|z = 1)$ and all elements of Γ_Y, with mixing probabilities $P(z = 1)$ and $P(z = 0)$. The identification region is a proper subset of Γ_Y whenever the probability $P(z = 0)$ of missing data is less

6

than one, and is a singleton when $P(z = 0) = 0$. Hence $P(y)$ is *partially identified* when $0 < P(z = 0) < 1$ and is *point-identified* when $P(z = 0) = 0$.

Distributional assumptions may have identifying power. One may assert that the distribution $P(y|z = 0)$ of missing outcomes lies in some set $\Gamma_{0Y} \subset \Gamma_Y$. Then the identification region shrinks from $H[P(y)]$ to

$$H_1[P(y)] \equiv [P(y|z = 1)P(z = 1) + \gamma P(z = 0), \gamma \in \Gamma_{0Y}]. \qquad (1.3)$$

Or one may assert that the distribution of interest, $P(y)$, lies in some set $H_0[P(y)] \subset \Gamma_Y$. Then the identification region shrinks from $H[P(y)]$ to

$$H_1[P(y)] \equiv H_0[P(y)] \cap H[P(y)]. \qquad (1.4)$$

Assumptions of the former and latter types differ in that the former are necessarily non-refutable but the latter may be refutable. An assumption that restricts $P(y|z = 0)$ is non-refutable because, after all, one does not observe the missing data. In contrast, an assumption that restricts $P(y)$ may be incompatible with the available empirical evidence. If the intersection of $H_0[P(y)]$ and $H[P(y)]$ is empty, one should conclude that $P(y)$ does not lie in the set $H_0[P(y)]$.

The above concerns identification of the entire outcome distribution. A common objective of empirical research is to infer a parameter of this distribution; for example, one may want to learn the mean of y. Viewing this in abstraction, let $\tau(\cdot): \Gamma_Y \rightarrow T$ be a function mapping probability distributions on Y into a space T and consider the problem of inference on the parameter $\tau[P(y)]$. The identification region for $\tau[P(y)]$ is

$$H\{\tau[P(y)]\} = \{\tau(\eta), \eta \in H[P(y)]\} \qquad (1.5)$$

if only the empirical evidence is available and is

$$H_1\{\tau[P(y)]\} = \{\tau(\eta), \eta \in H_1[P(y)]\} \qquad (1.6)$$

given distributional assumptions as discussed above.

Statistical Inference

The fundamental problem posed by missing data is identification, so it is analytically convenient to suppose that one knows the distributions that are asymptotically revealed by the sampling process, namely $P(y|z = 1)$ and $P(z)$. Of course, an empirical researcher observing a sample of finite size N must contend with issues of statistical inference as well as identification. I shall not dwell on these here, but merely point out that the empirical distributions $P_N(y|z=1)$ and $P_N(z)$ almost surely converge to $P(y|z=1)$ and $P(z)$ respectively. Hence, a natural nonparametric estimate of the identification region $H[P(y)]$ is the sample analog

$$H_N[P(y)] \equiv [P_N(y|z=1)P_N(z=1) + \gamma P_N(z=0), \gamma \in \Gamma_Y] \qquad (1.7)$$

and a natural nonparametric estimate of $\{\tau(\eta), \eta \in H[P(y)]\}$ is $\{\tau(\eta), \eta \in H_N[P(y)]\}$. Sample analogs may also be used to estimate $H_1[P(y)]$ in the presence of distributional assumptions.

The Task Ahead

The above, in a nutshell, is the story of identification when the data are generated by random sampling and some outcome realizations are not observable. The task ahead is to flesh out and elaborate on this story.

The remainder of this chapter studies identification using the empirical evidence alone. Sections 1.2 and 1.3 describe the identification regions for particular parameters of interest: means of real-valued functions of y and parameters that respect stochastic dominance. Section 1.4 generalizes the premise of random sampling to cases in which data are available from multiple sampling processes, each process drawing persons at random and each having some missing data. Section 1.5 extends the scope of the analysis from missing data to interval measurement of outcomes.

Chapter 2 examines a broad class of distributional assumptions that use instrumental variables to help identify the distribution of outcomes. One such assumption is the long-familiar supposition that data are missing at random. Another is the premise that outcomes are statistically independent of an instrumental variable. Yet another is that mean outcomes vary monotonically with an instrumental variable.

The analysis here and in Chapter 2 extends immediately to inference on conditional outcome distributions if the conditioning event is always observed. One simply needs to redefine the population of interest to be the sub-population for which the conditioning event holds. Chapter 3 examines inference on conditional distributions when data on outcomes and/or conditioning events may be missing.

1.2. Means

Let $R \equiv [-\infty, \infty]$ be the extended real line. Let G be the space of measurable functions that map Y into R and that attain their lower and upper bounds g_0

$\equiv \inf_{y \in Y} g(y)$ and $g_1 \equiv \sup_{y \in Y} g(y)$. Thus $g \in G$ if there exists a $y_{0g} \in Y$ such that $g(y_{0g}) = g_0$ and a $y_{1g} \in Y$ such that $g(y_{1g}) = g_1$. The lower bound g_0 may be finite or may be $-\infty$; similarly, g_1 may be finite or may be ∞.

Let the problem of interest be to infer the expectation $E[g(y)]$ using only the empirical evidence. The Law of Iterated Expectations gives

$$E[g(y)] = E[g(y)|z=1]P(z=1) + E[g(y)|z=0]P(z=0). \qquad (1.8)$$

The sampling process asymptotically reveals $E[g(y)|z=1]$ and $P(z)$. However, it is uninformative regarding $E[g(y)|z=0]$, which can take any value in the interval $[g_0, g_1]$. Hence we have this simple, important result:

Proposition 1.1: Let $g \in G$. Given the empirical evidence alone, the identification region for $E[g(y)]$ is the closed interval

$$H\{E[g(y)]\} = [E[g(y)|z=1]P(z=1) + g_0 P(z=0),$$

$$E[g(y)|z=1]P(z=1) + g_1 P(z=0)]. \qquad (1.9)$$

□

If the function g does not attain its lower (upper) bound on Y, Proposition 1.1 remains valid with the closed interval on the right side of (1.9) replaced by one that is open from below (above).

Observe that $H\{E[g(y)]\}$ is a proper subset of $[g_0, g_1]$, and hence informative, whenever the probability $P(z=0)$ of missing data is less than one and g has finite range. The width of the region is $(g_1 - g_0)P(z=0)$. Thus, the severity of the identification problem varies directly with the probability $P(z=0)$ of missing data.

The situation changes if $g_0 = -\infty$ or $g_1 = \infty$. The identification region is the tail interval $[-\infty, E[g(y)|z=1]P(z=1) + g_1 P(z=0)]$ in the former case and $[E[g(y)|z=1]P(z=1) + g_0 P(z=0), \infty]$ in the latter. In both cases, the region remains informative but has infinite length. The region is $[-\infty, \infty]$ if g is unbounded from both below and above. Thus, credible prior information is a prerequisite for inference on the mean of an unbounded random variable.

Probabilities of Events

Proposition 1.1 has many applications. Perhaps the most far-reaching is the identification region it implies for the probability that y lies in any non-empty, proper, measurable set $B \subset Y$. Let $g_B(\cdot)$ be the indicator function

$g_B(y) \equiv 1[y \in B]$; that is, $g_B(y) = 1$ if $y \in B$ and $g_B(y) = 0$ otherwise. Then $g_B(\cdot)$ attains its lower and upper bounds on Y, these being 0 and 1. Moreover, $E[g_B(y)] = P(y \in B)$ and $E[g_B(y)|z=1] = P(y \in B|z=1)$. Hence, Proposition 1.1 has this corollary.

Corollary 1.1.1: Let B be a non-empty, proper, and measurable subset of Y. Given the empirical evidence alone, the identification region for $P(y \in B)$ is the closed interval

$$H[P(y \in B)] = [P(y \in B|z=1)P(z=1),$$

$$P(y \in B|z=1)P(z=1) + P(z=0)]. \qquad (1.10)$$

$\square$

The width of this interval is $P(z = 0)$, whatever the set B may be. The location of the interval does vary with B. In particular, if $B' \subset B$, the interval $H[P(y \in B)]$ shifts the interval $H[P(y \in B')]$ rightward.

Statistical Inference

The natural nonparametric estimate of the identification region for $E[g(y)]$ is its sample analog. The sampling distribution of this estimate is particularly simple to analyze if one rewrites (1.9) in the alternative form

$$H\{E[g(y)]\} = [E[g(y)z + g_0(1 - z)], E[g(y)z + g_1(1 - z)]]. \qquad (1.9')$$

The sample analog of (1.9') is the interval

$$H_N\{E[g(y)]\} = [E_N[g(y)z + g_0(1 - z)], E_N[g(y)z + g_1(1 - z)]] \qquad (1.11)$$

connecting the sample averages of $g(y)z + g_0(1 - z)$ and $g(y)z + g_1(1 - z)$. Hence, analysis of the sampling distribution of $H_N\{E[g(y)]\}$ is the elementary problem of analysis of the sampling distribution of a bivariate sample average.

Point Estimation and Tests of Hypotheses

Empirical researchers confronting missing data routinely combine the empirical evidence with distributional assumptions to produce point estimates of parameters of interest. These assumptions may or may not be correct, so it is of interest to examine point estimation from the present nonparametric perspective.

Let θ_N denote a point estimate of $E[g(y)]$ obtained in a sample of size N, and let θ be its probability limit. Suppose that θ lies outside the identification region $H\{E[g(y)]\}$. Then $E[g(y)]$ cannot equal θ. Hence, the asserted assumptions must be incorrect.

Of course, such a definitive conclusion cannot be drawn from finite-sample data. However, one can compare the point estimate θ_N with the identification-region estimate $H_N\{E[g(y)]\}$. This suggests statistical tests of the form: Reject the asserted assumptions if the point θ_N is sufficiently distant from the interval $H_N\{E[g(y)]\}$.

Such tests are applicable only when assumptions are refutable, in the sense that θ can logically lie outside the identification region $H\{E[g(y)]\}$. Chapter 2 studies a variety of assumptions, of which some are refutable and others are not.[1]

1.3. Parameters that Respect Stochastic Dominance

This section generalizes Proposition 1.1 from means of functions of y to parameters that respect stochastic dominance.

Parameters that Respect Stochastic Dominance (D-parameters): Let Γ_R be the space of probability distributions on the extended real line R. Distribution $F \in \Gamma_R$ stochastically dominates distribution $F' \in \Gamma_R$ if $F[-\infty, t] \leq F'[-\infty, t]$ for all $t \in R$. An extended real-valued function $D(\cdot): \Gamma_R \to R$ respects stochastic dominance (is a D-parameter) if $D(F) \geq D(F')$ whenever F stochastically dominates F'.

Leading examples of D-parameters are the mean and quantiles of real random variables. Spread parameters such as the variance or interquartile range do not respect stochastic dominance.

Here is the result:

Proposition 1.2: Let $D(\cdot)$ respect stochastic dominance. Let $g \in G$. Let $R_g \equiv [g(y), y \in Y]$ be the range set of g. Let Γ_g be the space of probability distributions on R_g. Let $\gamma_{0g} \in \Gamma_g$ and $\gamma_{1g} \in \Gamma_g$ be the degenerate distributions that place all mass on g_0 and g_1 respectively. Given the empirical evidence alone, the smallest and largest points in the identification region for $D\{P[g(y)]\}$ are $D\{P[g(y)|z = 1]P(z = 1) + \gamma_{0g}P(z = 0)\}$ and $D\{P[g(y)|z = 1]P(z = 1) + \gamma_{1g}P(z = 0)\}$. $\square$

Proof: The identification region for distribution $P[g(y)]$ is

$$H\{P[g(y)]\} \equiv \{P[g(y)|z=1]P(z=1) + \gamma P(z=0), \gamma \in \Gamma_g\}. \quad (1.12)$$

Consider distribution $P[g(y)|z=1]P(z=1) + \gamma_{0g}P(z=0)$, which supposes that all missing data take a value y_{0g} that minimizes g. This distribution belongs to $H\{P[g(y)]\}$ and is stochastically dominated by all other members of $H\{P[g(y)]\}$. Similarly, $P[g(y)|z=1]P(z=1) + \gamma_{1g}P(z=0)$ belongs to $H\{P[g(y)]\}$ and stochastically dominates all other members of $H\{P[g(y)]\}$. The result follows.

Q. E. D.

Proposition 1.2 determines sharp lower and upper bounds on $D\{P[g(y)]\}$, but it does not assert that the identification region is the entire interval connecting these bounds. Proposition 1.1 showed that the identification region is this interval if D is the expectation parameter. However, the interval may contain non-feasible values if D is another parameter that respects stochastic dominance.

A particularly simple example occurs when $g(y)$ is a binary random variable and D is a quantile of $P[g(y)]$. A quantile must be an element of the range set R_g. Hence, $D\{P[g(y)]\}$ cannot take a value in the interior of the interval $[0, 1]$.

Quantiles

Quantiles are familiar parameters that respect stochastic dominance. For $\alpha \in (0, 1)$, the α–quantile of $P[g(y)]$ is $Q_\alpha[g(y)] \equiv \min t: \{P[g(y) \le t] \ge \alpha\}$. Proposition 1.2 shows that the smallest feasible value of $Q_\alpha[g(y)]$ is the α–quantile of distribution $P[g(y)|z=1]P(z=1) + \gamma_{0g}P(z=0)$ and the largest feasible value is the α–quantile of $P[g(y)|z=1]P(z=1) + \gamma_{1g}P(z=0)$. Examination of these quantities yields the following Corollary.

Corollary 1.2.1: Let $\alpha \in (0,1)$. Define $r(\alpha)$ and $s(\alpha)$ as follows:

$r(\alpha) \equiv [1 - (1 - \alpha)/P(z=1)]$–quantile of $P[g(y)|z=1]$ if $P(z=1) > 1 - \alpha$
$\quad\quad \equiv g_0$ otherwise.

$s(\alpha) \equiv [\alpha/P(z=1)]$–quantile of $P[g(y)|z=1]$ if $P(z=1) \ge \alpha$
$\quad\quad \equiv g_1$ otherwise.

The smallest and largest points in the identification region for $Q_\alpha[g(y)]$ are $r(\alpha)$ and $s(\alpha)$.   □

Observe that $r(\alpha)$ and $s(\alpha)$ are weakly increasing functions of α; hence, the identification region for $Q_\alpha[g(y)]$ shifts to the right as α increases.

For any value of α, the lower and upper bounds $r(\alpha)$ and $s(\alpha)$ are generically informative if $P(z = 1) > 1 - \alpha$ and $P(z = 1) \geq \alpha$, respectively. This holds whether or not the function g has finite range. Clearly, the implications of missing data for inference on quantiles are quite different from the implications for inference on means.

Outer Bounds on Differences between D-Parameters

Sometimes the parameter of interest is the difference between two specified D-parameters; that is, a parameter of the form $\tau_{21}\{P[g(y)]\} \equiv D_2\{P[g(y)]\} - D_1\{P[g(y)]\}$. For example, the interquartile range $Q_{0.75}[g(y)] - Q_{0.25}[g(y)]$ is a familiar measure of the spread of a distribution. The mean–median difference $E[g(y)] - Q_{0.5}[g(y)]$ measures skewness.

In general, differences between D-parameters are not themselves D-parameters. Nevertheless, Proposition 1.2 may be used to obtain informative *outer bounds* on such differences. A lower bound on $\tau_{21}\{P[g(y)]\}$ is the proposition's lower bound on $D_2\{P[g(y)]\}$ minus its upper bound on $D_1\{P[g(y)]\}$; similarly, an upper bound on $\tau_{21}\{P[g(y)]\}$ is the proposition's upper bound on $D_2\{P[g(y)]\}$ minus its lower bound on $D_1\{P[g(y)]\}$.

The bound on $\tau_{21}\{P[g(y)]\}$ obtained in this manner generally is non-sharp; hence the term *outer bound*. Consider the lower bound constructed above for $\tau_{21}\{P[g(y)]\}$. For this to be sharp, there would have to exist a distribution of missing data that jointly makes $D_2\{P[g(y)]\}$ attain its sharp lower bound and $D_1\{P[g(y)]\}$ attain its sharp upper bound. However, $D_2\{P[g(y)]\}$ attains its sharp lower bound if all missing data take a value y_{0g} that minimizes g, and $D_1\{P[g(y)]\}$ attains its sharp upper bound if all missing data take a value y_{1g} that maximizes g. These two requirements are compatible with one another only in degenerate cases.

1.4. Combining Multiple Sampling Processes

I have so far presumed that the available data are generated by random sampling, with y observable if $z = 1$. This section generalizes the analysis to cases in which data from multiple sampling processes are available. Each sampling process draws persons at random from population J, and each yields some observable outcomes. Outcomes that are observable under some sampling processes may be missing under others. The objective is to combine the data generated by the sampling processes to learn as much as possible about $P(y)$.[2]

The possibility of combining data from multiple sampling processes arises often in survey research. One survey of a population of interest may attempt to interview respondents face-to-face, another by telephone, and another by mail or e-mail. Each interview mode may yield its own pattern of nonresponse.

Identification of $P(y)$

Let M denote the set of sampling processes. For each $j \in J$ and $m \in M$, let $z_{jm} = 1$ if y_j is observable under sampling process m; let $z_{jm} = 0$ otherwise. For each $m \in M$, the Law of Total Probability gives

$$P(y) = P(y|z_m = 1)P(z_m = 1) + P(y|z_m = 0)P(z_m = 0). \tag{1.13}$$

Sampling process m asymptotically reveals the distribution of observable outcomes, $P(y|z_m = 1)$, and the distribution of observability, $P(z_m)$. Hence

$$P(y) \in [P(y|z_m = 1)P(z_m = 1) + \gamma_m P(z_m = 0), \ \gamma_m \in \Gamma_Y]. \tag{1.14}$$

The set of sampling processes collectively reveal that $P(y)$ lies in the intersection of the sets of distributions on the right side of (1.14). Hence, we have the following proposition.

Proposition 1.3: The identification region for $P(y)$ is

$$H_M[P(y)] \equiv \bigcap_{m \in M} [P(y|z_m = 1)P(z_m = 1) + \gamma_m P(z_m = 0), \ \gamma_m \in \Gamma_Y]. \tag{1.15}$$

$\square$

Proposition 1.3 is simple in form but is too abstract to communicate much about the size and shape of the identification region. Corollary 1.3.1 gives a useful alternative characterization of the region when $|M|$ is finite. Part (a) shows that a distribution is a feasible value for $P(y)$ if and only if the probability that it places on each measurable subset of Y is no less than an easily computed lower bound. This characterization further simplifies when Y is countable. Then, part (b) shows that one need only consider the probability placed on each atom of Y. This finding yields a simple necessary and sufficient condition for existence of a unique feasible distribution, given in part (c).

Corollary 1.3.1: Let $H_M[P(y)]$ be given by (1.15), with $|M|$ finite. Let $\eta \in \Gamma_Y$. For each measurable set $B \subset Y$, define

$$\pi_M(B) \equiv \max_{m \in M} P(y \in B | z_m = 1) P(z_m = 1). \qquad (1.16)$$

(a) Then $\eta \in H_M[P(y)]$ if and only if $\eta(B) \geq \pi_M(B)$, $\forall\, B \subset Y$.

(b) Let Y be countable. Then $\eta \in H_M[P(y)]$ if and only if $\eta(y) \geq \pi_M(y)$, $\forall\, y \in Y$.

(c) Let Y be countable. Let $S_M \equiv \sum_{y \in Y} \pi_M(y)$. The region $H_M[P(y)]$ contains multiple distributions if $S_M < 1$ and a unique distribution if $S_M = 1$. If $S_M = 1$, the unique feasible distribution is $\eta_M(y) \equiv \pi_M(y)$, $y \in Y$. $\qquad\square$

Proof: (a) Let $\eta \in H_M[P(y)]$. For each $m \in M$, there exists a distribution $\gamma_m \in \Gamma_Y$ such that $\eta = P(y | z_m = 1) P(z_m = 1) + \gamma_m P(z_m = 0)$. Hence $\eta(B) \geq P(y \in B | z_m = 1) P(z_m = 1)$, $\forall\, B \subset Y$. Hence $\eta(B) \geq \pi_M(B)$, $\forall\, B \subset Y$.

Let $\eta(B) \geq \pi_M(B)$, $B \subset Y$. Let $\gamma_m \equiv [\eta - P(y|z_m = 1) P(z_m = 1)]/P(z_m = 0)$. Then γ_m is a probability measure. Moreover, $\eta = P(y | z_m = 1) P(z_m = 1) + \gamma_m P(z_m = 0)$. Hence $\eta \in H_M[P(y)]$.

(b) Part (a) shows directly that $\eta \in H_M[P(y)] \Rightarrow \eta(y) \geq \pi_M(y)$, $\forall\, y \in Y$. Let $\eta(y) \geq \pi_M(y)$, $y \in Y$. Then for all $B \subset Y$,

$$\eta(B) = \sum_{y \in B} \eta(y) \geq \sum_{y \in B} \pi_M(y) \geq \pi_M(B), \qquad (1.17)$$

where the final inequality holds because $\pi_M(\cdot)$ is sub-additive. Hence $\eta \in H_M[P(y)]$, again by part (a).

(c) If $S_M < 1$, the empirical evidence leaves indeterminate the allocation of $(1 - S_M)$ of probability mass among the atoms of Y; Hence $H_M[P(y)]$ contains multiple distributions.

If $S_M = 1$, then η_M is a probability measure. Part (b) shows that $\eta_M \in H_M[P(y)]$. Let η be a measure with $\eta(y) \geq \pi_M(y)$, $y \in Y$, and $\eta(y) > \pi_M(y)$ for some $y \in Y$. Then $\eta(Y) > 1$, so η is not a probability measure. Hence η_M is the only element of $H_M[P(y)]$.

Note that η_M is distinct from π_M, which is sub-additive and hence not a probability distribution. That is, $\eta_M(y) = \pi_M(y)$ for $y \in Y$ but $\eta_M(B) \geq \pi_M(B)$ for $B \subset Y$.

$\qquad\qquad\qquad\qquad\qquad\qquad\qquad\qquad\qquad\qquad\qquad$ Q. E. D.

Corollary 1.3.1 shows that when Y is countable, a sufficient statistic for $H_M[P(y)]$ is the vector $[\pi_M(y), y \in Y]$. We see immediately that $H_M[P(y)]$ shrinks as $[\pi_M(y), y \in Y]$ increases. Moreover, we can measure the size of $H_M[P(y)]$. As observed in the proof to part (c), the empirical evidence leaves indeterminate the allocation of $(1 - S_M)$ of probability mass among the atoms of Y, where $S_M \equiv \sum_{y \in Y} \pi_M(y)$. Hence, the size of $H_M[P(y)]$ as measured by the sup norm is

$$\|H_M\|_{sup} = \sup_{(\eta, \eta') \in H_M \times H_M} \sup_{B \subset Y} |\eta(B) - \eta'(B)| = 1 - S_M. \qquad (1.18)$$

For given values of $[P(y|z_m = 1), m \in M]$, S_M rises as the vector $[P(z_m = 1), m \in M]$ increases. Some insight into the role of $[P(y|z_m = 1), m \in M]$ in the determination of S_M may be obtained from the inequality

$$S_M \equiv \sum_{y \in Y} \max_{m \in M} P(y = y | z_m = 1) P(z_m = 1)$$

$$\geq \max_{m \in M} [\sum_{y \in Y} P(y = y | z_m = 1)] P(z_m = 1)$$

$$= \max_{m \in M} P(z_m = 1). \qquad (1.19)$$

The lower bound on S_M in (1.19) is attained if the distributions $P(y|z_m = 1)$, $m \in M$ are identical to one another; then $H_M[P(y)] = H_{m*}[P(y)]$, where $m* \equiv \text{argmax}_{m \in M} P(z_m = 1)$. Thus, combining multiple sampling processes is least informative when all sampling processes generate the same observed distribution of y.

The event $S_M > 1$ cannot occur if all sampling processes draw realizations at random from population J. If $S_M > 1$, a measure η that satisfies part (b) has $\eta(Y) > 1$ and so is not a probability measure. Hence $H_M[P(y)]$ is empty. If one finds that $S_M > 1$, one should conclude that some sampling process does not draw realizations at random from J.

Parameters that Respect Stochastic Dominance

When Y is a countable subset of the real line, Corollary 1.3.1 implies simple characterizations of the identification regions for parameters that respect stochastic dominance. Corollary 1.3.2 gives the result. Part (a) determines the endpoints of the identification region for any D-parameter. Part (b) focuses on the important special case of the expectation parameter and shows that its identification region is the closed interval connecting the endpoints determined in part (a).

Corollary 1.3.2: Let $H_M[P(y)]$ be given by (1.15), with $|M|$ finite. Let Y be a countable subset of R, and let Y contain its lower and upper bounds $y_0 \equiv \inf_{y \in Y}$ and $y_1 \equiv \sup_{y \in Y}$. Let η_0 and η_1 be probability distributions on Y such that, for each $y \in Y$,

$$\eta_0(y) = \pi_M(y) \text{ if } y > y_0 \text{ and } \eta_0(y_0) = \pi_M(y_0) + (1 - S_M), \qquad (1.20a)$$

$$\eta_1(y) = \pi_M(y) \text{ if } y < y_1 \text{ and } \eta_1(y_1) = \pi_M(y_1) + (1 - S_M). \qquad (1.20b)$$

(a) Let $D(\cdot)$ respect stochastic dominance. Then the smallest and largest elements of $H_M\{D[P(y)]\}$ are $D(\eta_0)$ and $D(\eta_1)$.

(b) The closed interval

$$H_M[E(y)] = [\textstyle\sum_{y \in Y} y\pi_M(y) + (1-S_M)y_0, \sum_{y \in Y} y\pi_M(y) + (1-S_M)y_1] \quad (1.21)$$

is the identification region for $E(y)$. $\square$

Proof:
   (a) Corollary 1.3.1 showed that $\eta \in H_M[P(y)]$ if and only if $\eta(y) \geq \pi_M(y)$, $\forall y \in Y$. By construction, η_0 and η_1 are members of $H_M[P(y)]$. Indeed, η_0 is stochastically dominated by all members of $H_M[P(y)]$, and η_1 stochastically dominates all members of $H_M[P(y)]$. Hence, the smallest and largest elements of $H_M\{D[P(y)]\}$ are $D(\eta_0)$ and $D(\eta_1)$.
   (b) The expectation parameter respects stochastic dominance. Hence, part (a) shows that the smallest and largest elements of $H_M[E(y)]$ are $\int y dP\eta_0$ and $\int y dP\eta_1$, which equal the endpoints of the interval on the right side of (1.21). For any $\delta \in [0, 1]$, the mixture $\delta\eta_0 + (1 - \delta)\eta_1$ belongs to $H_M[P(y)]$. Hence $H_M[E(y)]$ is the entire interval on the right side of (1.21).
                                                                 Q. E. D.

1.5. Interval Measurement of Outcomes

The phenomenon of missing outcomes juxtaposes extreme observational states: each realization of y is either observed completely or not observed at all. Empirical researchers sometimes encounter intermediate observational states, in which realizations of y are observed to lie in proper but non-unitary subsets of the outcome space Y. A particularly common intermediate observational state is interval measurement of real-valued outcomes.
   To formalize interval measurement, let $Y \subset R$. Let each $j \in J$ have a triple $(y_{j-}, y_j, y_{j+}) \in Y^3$. Let the random variable $(y_-, y, y_+): J \to Y^3$ have a distribution $P(y_-, y, y_+)$ such that

$$P(y_- \leq y \leq y_+) = 1. \qquad (1.22)$$

Let a sampling process draw persons at random from J. Then we have interval measurement of outcomes if realizations of (y_-, y_+) are observable but realizations of y are not directly observable. (I say that y is not "directly" observable to cover the possibility that $y_- = y_+$, in which case observation of (y_-, y_+) implies observation of y.)

Sampling with missing outcomes is a special case of interval measurement. Let $y_0 \equiv \inf_{y \in Y}$ and $y_1 \equiv \sup_{y \in Y}$. A realization of y is effectively observed when $(y_- = y_+)$ and missing when $(y_- = y_0, y_+ = y_1)$. Hence, sampling with missing outcomes is the special case of (1.22) in which

$$P(y_- = y_+) + P(y_- = y_0, y_+ = y_1) = 1. \tag{1.23}$$

Interval measurement of outcomes yields very simple lower and upper bounds on any parameter that respects stochastic dominance. The distribution $P(y_+)$ is a feasible value of $P(y)$ and stochastically dominates all other feasible values of $P(y)$; Hence $D[P(y_+)]$ is the largest feasible value of $D[P(y)]$. The distribution $P(y_-)$ is a feasible value of $P(y)$ and is stochastically dominated by all other feasible values of $P(y)$; Hence $D[P(y_-)]$ is the smallest feasible value of $D[P(y)]$. Thus, we have Proposition 1.4.

Proposition 1.4: Let $Y \subset R$, let (y_-, y_+) be observable, and let (1.22) hold. Let D respect stochastic dominance. Given the empirical evidence alone, the smallest and largest points in the identification region for $D[P(y)]$ are $D[P(y_-)]$ and $D[P(y_+)]$. $\square$

Complement 1A. Employment Probabilities

This complement presents an empirical example illustrating Corollary 1.1.1. Horowitz and Manski (1998) used data from the National Longitudinal Survey of Youth (NLSY) to estimate the probability that a member of the surveyed population is employed in 1991. The surveyed population consists of persons born between January 1, 1957 and December 31, 1964 who resided in the United States in 1979. From 1979 on, the NLSY has periodically sought to interview a random sample of this population and supplemental samples of some sub-populations (Center for Human Resource Research, 1992). The random-sample data are used here.

In this illustration, the outcome y indicates an individual's employment status at the time of the 1991 interview. In the 1979 base year, the NLSY sought to interview a random sample of 6812 individuals and succeeded in obtaining interviews from 6111 of the sample members. Data on employment status in 1991 are available for 5556 of the 6111 individuals interviewed in the base year. The remaining 555 are nonrespondents, some because they declined to be interviewed in 1991 and some because they did not answer the employment-status question in their 1991 interviews. Table 1.1 presents these response statistics and the frequencies with which different outcome values are reported.

Table 1.1: 1991 Employment Status of NLSY Respondents

Employment Status	Number of Respondents
Employed ($y = 2$)	4332
Unemployed ($y = 1$)	297
Out of Labor Force ($y = 0$)	927
Ever-interviewed Nonrespondents	555
Never-interviewed Nonrespondents	701
Total	6812

The discussion below first uses these frequencies to generate empirical probabilities of events and then interprets these empirical probabilities as finite-sample estimates of population quantities.

The empirical nonresponse rate, which takes account of sample members who were never interviewed, is $P(z = 0) = 1256/6812 = 0.184$. Researchers computing nonresponse rates to questions in the later years of longitudinal surveys often condition on the event that a sample member was interviewed in the base year. Let this event, which is always observed, be denoted BASE. Then, the "ever-interviewed" nonresponse rate for employment status in 1991 is $P(z = 0 | BASE) = 555/6111 = 0.091$.

The empirical probability of employment for the 5556 individuals who responded to the 1991 employment-status question is $P(y = 2 | z = 1) = 4332/5556 = 0.780$. The probability of employment among nonrespondents can take any value in the interval [0, 1]. Hence, Corollary 1.1.1 yields the following identification regions for the population and the ever-interviewed empirical employment probabilities $P(y = 2)$ and $P(y = 2 | BASE)$:

$$P(y = 2) \in [(0.780)(0.816), \ (0.780)(0.816) + (0.184)] = [0.636, 0.820],$$

$$P(y = 2 | BASE) \in [(0.780)(0.909), \ (0.780)(0.909) + (0.091)]$$

$$= [0.709, 0.800].$$

Sampling Variation

The focus of this book is identification, but empirical research must also be concerned with sampling variation. Thus, let us now consider the empirical probabilities analyzed above to be random sample estimates of corresponding population quantities. Then the effects of sampling variation may be characterized by confidence intervals on the identification regions obtained above.

Horowitz and Manski (1998) presented Bonferroni intervals based on local asymptotic theory.[3] Consider $P(y=2)$. The identification region (1.10) may be rewritten as

$$H[P(y=2)] \;=\; [P(y=2, z=1), \; 1 - P(y \neq 2, z=1)].$$

The asymptotic standard errors of the sample estimates of the lower and upper bounds of this interval are

$$
\begin{aligned}
C_L &= \{P(y=2, z=1)[1 - P(y=2, z=1)]/N\}^{1/2}, \\
C_U &= \{P(y \neq 2, z=1)[1 - P(y \neq 2, z=1)]/N\}^{1/2},
\end{aligned}
$$

where $N = 6812$ is the sample size. A Bonferroni asymptotic joint confidence region with level at least 95 percent is obtained by forming the intersection of individual 97.5 percent regions. These regions are the point estimates of the lower and upper bounds $\pm$ (2.24)C_L and $\pm$ (2.24)C_U respectively.

Substituting sample frequencies for population probabilities yields

$$
\begin{aligned}
P(y=2, z=1) &= 4332/6812 \;= 0.636, \\
P(y \neq 2, z=1) &= 1224/6812 \;= 0.180,
\end{aligned}
$$

$$
\begin{aligned}
C_L &= [(0.636)(1 - 0.636)(1/6812)]^{1/2} = 0.0058, \\
C_U &= [(0.180)(1 - 0.180)(1/6812)]^{1/2} = 0.0047,
\end{aligned}
$$

so the estimated asymptotic joint Bonferroni 95 percent intervals are

$$
\begin{aligned}
0.623 &\leq \text{lower bound on } P(y=2) \leq 0.649, \\
0.810 &\leq \text{upper bound on } P(y=2) \leq 0.831.
\end{aligned}
$$

Analogous computations conditioning on the event BASE yield

$$
\begin{aligned}
0.696 &\leq \text{lower bound on } P(y=2 \,|\, \text{BASE}) \leq 0.722, \\
0.788 &\leq \text{upper bound on } P(y=2 \,|\, \text{BASE}) \leq 0.811.
\end{aligned}
$$

These confidence intervals are much narrower than the widths of the identification regions. Thus, identification is the dominant problem in inference on $P(y = 2)$ and $P(y = 2 \mid \text{BASE})$ from the NLSY data; sampling variation is a second-order concern. This conclusion holds except when the sample size is quite small or the response rate is very close to one.

Complement 1B. Blind-Men Bounds on an Elephant

The ancient Indian fable *The Blind Men and the Elephant* exemplifies the problem of combining empirical evidence from multiple sampling processes, each of which partially identifies a population distribution of interest. Modern renditions of the fable agree on the inferential problem but differ on its resolution.

The nineteenth-century American poem of John Godfrey Saxe, reproduced here, concludes pessimistically; the six disputatious blind men fail to recognize that each has observed a different feature of the elephant. In other renditions, the blind men learn to combine their observations for their common benefit. For example, in a version of the fable prepared for classroom use, the blind men are interrupted by a listening Rajah who counsels as follows: "The elephant is a very large animal. Each man touched only one part. Perhaps if you put the parts together, you will see the truth." [4]

The Rajah may have been optimistic in suggesting that six partial observations of an elephant may reveal the full truth about the creature, but he was sensible to counsel that six partial observations are more informative than one. All too often, researchers act like the disputatious blind men of the Saxe poem, each failing to recognize that he or she has observed a different feature of the same population. They should instead combine their observations, as counseled by the Rajah.

"The Blind Men and the Elephant", by
John Godfrey Saxe (1816–1887)

It was six men of Indostan
To learning much inclined,
Who went to see the Elephant
(Though all of them were blind),
That each by observation
Might satisfy his mind

The First approached the Elephant,
And happening to fall
Against his broad and sturdy side,
At once began to bawl:
"God bless me! but the Elephant
Is very like a wall!"

The Second, feeling of the tusk,
Cried, "Ho! what have we here
So very round and smooth and sharp?
To me 'tis mighty clear
This wonder of an Elephant
Is very like a spear!"

The Third approached the animal,
And happening to take
The squirming trunk within his hands,
Thus boldly up and spake:
"I see," quoth he, "the Elephant
Is very like a snake!"

The Fourth reached out an eager hand,
And felt about the knee.
"What most this wondrous beast is like
Is mighty plain," quoth he;
" 'Tis clear enough the Elephant
Is very like a tree!"

The Fifth, who chanced to touch the ear,
Said: "E'en the blindest man
Can tell what this resembles most;
Deny the fact who can
This marvel of an Elephant
Is very like a fan!?

The Sixth no sooner had begun
About the beast to grope,
Than, seizing on the swinging tail
That fell within his scope,
"I see," quoth he, "the Elephant
Is very like a rope!"

And so these men of Indostan
Disputed loud and long,
Each in his own opinion
Exceeding stiff and strong,
Though each was partly in the right,
And all were in the wrong!

Endnotes

Sources and Historical Notes

The basic ideas in Sections 1.1 and 1.2 were first presented in Manski (1989) and were developed more fully in Manski (1994). Corollary 1.2.1 was proved directly, without reference to D-parameters, in Manski (1994, Proposition 2). Proposition 1.3 and its corollaries re-interpret and extend results proved in Manski (2003). Proposition 1.4 is based on Manski and Tamer (2002, Proposition 1).

The class of parameters that respect stochastic dominance was introduced in Horowitz and Manski (1995) and studied further in Manski (1997a). Many partial identification results for means of random variables extend easily to this class of parameters, as shown throughout this book.

As described in Manski (1989, 1995), my study of inference with missing outcome data grew out of a specific inquiry by Irving Piliavin in the spring of 1987. Piliavin and his colleague Michael Sosin had interviewed a sample of 137 individuals who were homeless in Minneapolis in late December 1985. Six months later, they attempted to re-interview these respondents to measure an outcome of interest but succeeded in locating only 78. Piliavin told me that he felt it implausible to assume that nonresponse to the second survey was random. Nor was he comfortable making other assumptions about the nonresponse process. He asked whether it was possible to draw inferences without imposing such assumptions.

Fifty years ago, in a study of the statistical problems of the Kinsey report on sexual behavior, Cochran, Mosteller, and Tukey (1954, pp. 274–282) used essentially Corollary 1.1.1 to express the possible effects of nonresponse to the Kinsey survey. However, the subsequent literature on missing data in surveys did not pursue the idea, preferring instead to impose distributional assumptions that yield point identification (e.g., Little and Rubin, 1987). I learned of the Cochran, Mosteller, and Tukey work in the early 1990s, and wondered why the authors did not pursue the idea of inference using the empirical evidence alone. I found that Cochran (1977) had subsequently dismissed such inference as uninformative in practice. Using the symbol W_2 to denote the probability of missing data, he wrote (p. 362): "The limits are distressingly wide unless W_2 is very small." Cochran did not discuss what a researcher should do in the absence of credible assumptions that shrink these bounds.

Text Notes

1. Point estimates obtained by *imputation* of missing values are non-refutable using the empirical evidence alone. Imputation methods assign to each person with a missing realization of y some logically possible value, say y^*. This done, $E[g(y)]$ is estimated by the sample average

$$\theta_N = \frac{1}{N} \sum_{i=1}^{N} g(y_i)z_i + g(y_i^*)(1 - z_i).$$

By the Strong Law of Large Numbers, θ_N almost surely converges to

$$\theta \equiv E[g(y)\,|\,z = 1]\cdot P(z = 1) + E[g(y^*)\,|\,z = 0]\cdot P(z = 0).$$

This θ necessarily lies in $H\{E[g(y)]\}$ but does not necessarily equal $E[g(y)]$. The latter holds if and only if $E[g(y^*)\,|\,z = 0] = E[g(y)\,|\,z = 0]$.

2. Although there has been much research on inference combining multiple sampling processes, the problem of partial identification examined here has not previously been addressed. The statistical literature on *meta-analysis* has supposed that each sampling process independently point-identifies the distribution of interest; that said, the concern has been to combine the available data sources in a statistically efficient manner. Econometric research on *sample augmentation* has considered situations in which each sampling process incompletely identifies the distribution of interest, but combining multiple sampling processes with suitable assumptions achieves point identification; see, for example, Hsieh, Manski, and McFadden (1985) and Hirano, Imbens, Ridder, and Rubin (2001).

3. The problem of obtaining a joint confidence interval for a pair of lower and upper bounds is examined more fully in Horowitz and Manski (2000). There we consider construction of a confidence interval with known (asymptotic) probability of containing both the lower and the upper bound on a partially identified population parameter. We focus on intervals of the form $[L_N - z_{N\alpha}, U_N + z_{N\alpha}]$, where N is sample size and L_N and U_N are estimates of the lower and upper bounds L and U on the parameter of interest. The number $z_{N\alpha}$ is chosen so that $P(L_N - z_{N\alpha} \le L, U \le U_N + z_{N\alpha}) = 1 - \alpha$ asymptotically. One way of obtaining $z_{N\alpha}$ is to derive it from an analytic expression for the asymptotic distribution of (L_N, U_N). Another is to use the bootstrap.

4. A World Wide Web search on the phrase "The Blind Men and the Elephant" yields many versions of the fable. One with the Rajah's advice is at the URL www.peacecorps.gov/wws/guides/looking/story22.html.

2

Instrumental Variables

2.1. Distributional Assumptions and Credible Inference

Distributional assumptions may enable one to shrink identification regions obtained using empirical evidence alone. When facing the problem of missing outcome data, researchers have generally imposed distributional assumptions that point-identify the outcome distribution $P(y)$. When a single random sampling process generates the available data, it has been particularly common to assert that observed and missing outcomes have the same distribution; that is,

$$P(y) = P(y|z = 0) = P(y|z = 1). \qquad (2.1)$$

The distribution $P(y|z = 1)$ is revealed by the sampling process, so $P(y)$ is point-identified. Someone asserting (2.1) cannot be proved wrong; after all, the empirical evidence reveals nothing about $P(y|z = 0)$.

An assumption may be non-refutable and yet not credible. Researchers who assert (2.1) almost inevitably find this assumption difficult to justify. Analysts who assert other point-identifying assumptions regularly encounter the same difficulty. This should not be surprising. The empirical evidence reveals nothing at all about the distribution of missing data. An assumption must be quite strong to pick out one among all possible distributions.

There is a fundamental tension between the credibility and strength of conclusions, which I have called the *Law of Decreasing Credibility*. Inference using the empirical evidence alone sacrifices strength of conclusions in order to maximize credibility. Inference invoking point-identifying distributional assumptions sacrifices credibility in order to achieve strong conclusions. Between these poles, there is a vast middle ground of possible modes of inference asserting assumptions that may shrink the identification region $H[P(y)]$ but not reduce it to a point.

This chapter examines the identifying power of various distributional assumptions that make use of instrumental variables. Some such assumptions imply point identification, whereas others have less identifying power and, perhaps, greater credibility. For simplicity, the analysis below presumes that a single random sampling process generates the available data, that realizations of y are either completely observed or entirely missing, and that all realizations of the instrumental variable are observed. Distributional assumptions using instrumental variables may also be applied when data are available from multiple sampling processes, when interval measures of outcomes are observed, and when some realizations of the instrumental variable are missing.

2.2. Some Assumptions Using Instrumental Variables

As in Chapter 1, suppose that a sampling process draws persons at random from population J and that the outcome y is observable if $z = 1$. Moreover, suppose now that each person j is characterized by a covariate v_j in a space V. Let $v: J \to V$ be the random variable mapping persons into covariates and let $P(y, z, v)$ denote the joint distribution of (y, z, v). Suppose that all realizations of v are observable. Observability of v provides an instrument or tool that may help to identify the outcome distribution $P(y)$. Thus v is said to be an *instrumental variable*.

The sampling process asymptotically reveals the distributions $P(z)$, $P(y, v|z = 1)$, and $P(v|z = 0)$. It is uninformative about the conditional distributions $[P(y|v = v, z = 0), v \in V]$. The presence of an instrumental variable does not, per se, help to identify $P(y)$. However, observability of v may be useful when combined with distributional assumptions. This chapter examines the identifying power of six such assumptions.

Sections 2.3 and 2.4 study identification of $P(y)$ under assumptions that assert forms of statistical independence among the random variables (y, z, v). Section 2.3 assumes that observed and missing outcomes have the same distribution conditional on v; that is, outcomes are *missing-at-random* conditional on v.

Outcomes Missing-at-Random (Assumption MAR):

$$P(y|v) = P(y|v, z = 0) = P(y|v, z = 1). \qquad (2.2)$$

Section 2.4 assumes that y is statistically independent of v; that is:

Statistical Independence of Outcomes and Instruments (Assumption SI):

$$P(y|v) = P(y). \qquad (2.3)$$

Section 2.5 studies identification of the expectation $E[g(y)]$ of a real-valued function $g(y)$ under assumptions that are weaker than Assumptions MAR and SI. First the forms of statistical independence asserted in (2.2) and (2.3) are weakened to the mean-independence assumptions

Means Missing-at-Random (Assumption MMAR):

$$E[g(y)|v] = E[g(y)|v, z = 0] = E[g(y)|v, z = 1] \qquad (2.4)$$

and

Mean Independence of Outcomes and Instruments (Assumption MI):

$$E[g(y)|v] = E[g(y)], \qquad (2.5)$$

respectively. Then, Assumptions MMAR and MI are weakened to the monotonicity assumptions

Mean Missing Monotonically (Assumption MMM):

$$E[g(y)|v, z = 1] \geq E[g(y)|v] \geq E[g(y)|v, z = 0] \qquad (2.6)$$

and

Mean Monotonicity of Outcomes and Instruments (Assumption MM): Let V be an ordered set.

$$E[g(y)|v = v] \geq E[g(y)|v = v'], \forall (v, v') \in V \times V \text{ such that } v \geq v'. \ (2.7)$$

Taken together, these six distributional assumptions provide a variety of ways in which a researcher can use instrumental variables to help identify outcome distributions when some outcome data are missing. Researchers contemplating the use of instrumental variables should, of course, pay due attention to the credibility of these and other assumptions. Empirical researchers often ask whether some observable covariate is or is not a "valid instrument" in an application of interest. The expression "valid instrument" is imprecise because it focuses attention on the covariate used in the role of v. Credibility depends not on the covariate per se but on the assumption that the distribution $P(y, z, v)$ is assumed to satisfy.

To simplify the presentation, the analysis below supposes that the covariate space V is finite and that $P(v = v, z = 1) > 0$ for all $v \in V$. These regularity conditions are maintained without further reference.

2.3. Outcomes Missing-at-Random

Assumption MAR is a non-refutable hypothesis that point-identifies $P(y)$. Proposition 2.1 shows how.[1]

Proposition 2.1: Let assumption MAR hold. Then $P(y)$ is point-identified with

$$P(y) = \sum_{v \in V} P(y | v = v, z = 1) P(v = v). \qquad (2.8)$$

Assumption MAR is non-refutable.   □

Proof: The Law of Total Probability gives

$$P(y) = \sum_{v \in V} P(y | v = v) P(v = v). \qquad (2.9)$$

Assumption MAR states that

$$P(y | v) = P(y | v, z = 1). \qquad (2.10)$$

Applying (2.10) to (2.9) yields (2.8). The right side of (2.8) is point-identified by the sampling process, so $P(y)$ is point-identified. Assumption MAR is non-refutable because the empirical evidence reveals nothing about $P(y | v, z = 0)$.

Q. E. D.

A researcher applying assumption MAR must specify the instrumental variable v for which the assumption holds. Assumption (2.1) is the special case in which v has a degenerate distribution. As in that case, the credibility of assumption MAR is regularly a matter of controversy.[2]

2.4. Statistical Independence

Assumption SI has the same identifying power as does observation of data from multiple sampling processes. The space V of values for the instrumental variable plays the same role here as did the set M of sampling processes in Section 1.4. Proposition 2.2 gives the basic result, and two corollaries flesh it out.

Proposition 2.2: (a) Let assumption SI hold. Then the identification region for $P(y)$ is

$$H_{SI}[P(y)] \;=\;$$

$$\bigcap_{v \in V} \{P(y|v=\mathrm{v}, z=1)P(z=1|v=\mathrm{v}) + \gamma_\mathrm{v} \cdot P(z=0|v=\mathrm{v}), \ \gamma_\mathrm{v} \in \Gamma_Y\}.$$

$$(2.11)$$

(b) Let the set $H_{SI}[P(y)]$ be empty. Then assumption SI does not hold.   □

Proof: (a) Application of equation (1.2) to each conditional distribution $P(y|v=\mathrm{v})$, $\mathrm{v} \in V$ gives the identification region for this distribution using the empirical evidence alone; that is,

$$H[P(y|v=\mathrm{v})] \;=\;$$

$$[P(y|v=\mathrm{v}, z=1)P(z=1|v=\mathrm{v}) + \gamma_\mathrm{v} \cdot P(z=0|v=\mathrm{v}), \ \gamma_\mathrm{v} \in \Gamma_Y].$$

$$(2.12)$$

Moreover, the identification region for the set of distributions $[P(y|v=\mathrm{v}), \mathrm{v} \in V]$ is the Cartesian product $\times_{\mathrm{v} \in V} H[P(y|v=\mathrm{v})]$.

Assumption SI states that the distributions $P(y|v=\mathrm{v})$, $\mathrm{v} \in V$ coincide, all being equal to $P(y)$. Hence $P(y)$ must lie in $\cap_{\mathrm{v} \in V} H[P(y|v=\mathrm{v})]$. Any distribution in this intersection is feasible, so $H_{SI}[P(y)]$ is the identification region.

(b) If assumption SI holds, the set $H_{SI}[P(y)]$ is necessarily non-empty. Hence, the assumption cannot hold if $H_{SI}[P(y)]$ is empty.

Q. E. D.

Part (a) of the proposition shows that the identifying power of assumption SI can range from point identification of $P(y)$ to no power at all, depending on the nature of the instrumental variable. Point identification occurs if there exists a $v \in V$ such that $P(z = 1 | v = v)$; then one of the sets whose intersection is taken in (2.11) is a singleton. When Y is countable, Corollary 2.2.1 below gives a simple necessary and sufficient condition for point identification.

Assumption SI has no identifying power if (a) z is statistically independent of v and (b) y is statistically independent of v conditional on the event $\{z = 1\}$; that is, if $P(z|v) = P(z)$ and $P(y|v, z = 1) = P(y|z = 1)$. Then $H[P(y|v = v)]$, $v \in V$ are all the same as the identification region obtained using the empirical evidence alone. This shows that identification cannot be achieved by construction of a trivial instrumental variable that uses a randomization device to assign a covariate value to each member of the population. A covariate v generated by a randomization device is necessarily statistically independent of the pair (y, z). Such a covariate satisfies assumption SI but has no identifying power.

Part (b) of the proposition shows that assumption SI is refutable. If $H_{SI}[P(y)]$ is empty, the assumption logically cannot hold. Of course, non-emptiness of $H_{SI}[P(y)]$ does not imply that the assumption is correct.

Observe that the identification region $H_{SI}[P(y)]$ has the same structure as the region $H_M[P(y)]$ obtained by combining data from multiple sampling processes (see Proposition 1.3), with V here playing the role of M there. Hence, there are instrumental-variable analogs to Corollaries 1.3.1 and 1.3.2. These are given in Corollaries 2.2.1 and 2.2.2 below. The proofs are analogous to those of the earlier corollaries and so are omitted.

Corollary 2.2.1: Let assumption SI hold. Let $\eta \in \Gamma_Y$. For $B \subset Y$, define

$$\pi_V(B) \equiv \max_{v \in V} P(y \in B | v = v, z = 1) P(z = 1 | v = v). \qquad (2.13)$$

(a) Then $\eta \in H_{SI}[P(y)]$ if and only if $\eta(B) \geq \pi_V(B)$, $\forall B \subset Y$.

(b) Let Y be countable. Then $\eta \in H_{SI}[P(y)]$ if and only if $\eta(y) \geq \pi_V(y)$, $\forall y \in Y$.

(c) Let Y be countable. Let $S_V \equiv \sum_{y \in Y} \pi_V(y)$. Then $H_{SI}[P(y)]$ contains multiple distributions if $S_V < 1$ and a unique distribution if $S_V = 1$. If $S_V = 1$, the unique feasible distribution is $\eta_V(y) \equiv \pi_V(y)$, $y \in Y$. If $S_V > 1$, then assumption SI does not hold. $\qquad \square$

Corollary 2.2.2: Let assumption SI hold. Let Y be a countable subset of R, and let Y contain its lower and upper bounds $y_0 \equiv \inf_{y \in Y}$ and $y_1 \equiv \sup_{y \in Y}$. Let η_0 and η_1 be probability distributions on Y such that, for each $y \in Y$,

$$\eta_0(y) = \pi_V(y) \quad \text{if } y > y_0 \text{ and } \eta_0(y_0) = \pi_V(y_0) + (1 - S_V) \qquad (2.14a)$$

$$\eta_1(y) = \pi_V(y) \quad \text{if } y < y_1 \text{ and } \eta_1(y_1) = \pi_V(y_1) + (1 - S_V). \qquad (2.14b)$$

(a) Let $D(\cdot)$ respect stochastic dominance. Then the smallest and largest elements of $H_{SI}\{D[P(y)]\}$ are $D(\eta_0)$ and $D(\eta_1)$.

(b) The closed interval

$$H_{SI}[E(y)] = [\textstyle\sum_{y \in Y} y\pi_V(y) + (1 - S_V)y_0, \ \sum_{y \in Y} y\pi_V(y) + (1 - S_V)y_1]$$

$$(2.15)$$

is the identification region for $E(y)$. $\qquad\qquad\qquad\qquad\qquad\qquad\square$

2.5. Mean Independence and Mean Monotonicity

This section studies identification of the expectations of real-valued functions of the outcome. The distributional assumptions considered here are weaker than Assumptions MAR and SI. Throughout this section, $g(\cdot)$ is a real-valued function that attains its lower and upper bounds.

Mean Independence

Assumptions MMAR and MI weaken the forms of statistical independence asserted in assumptions MAR and SI to corresponding forms of mean independence. Assumption MMAR is a non-refutable hypothesis that point-identifies $E[g(y)]$. Assumption MI is a refutable hypothesis that generically shrinks the identification region obtained using the empirical evidence alone, but point-identifies $E[g(y)]$ only in special cases. Propositions 2.3 and 2.4 give the results.

Proposition 2.3: Let assumption MMAR hold. Then $E[g(y)]$ is point-identified with

$$E[g(y)] = \textstyle\sum_{v \in V} E[g(y)|v = v, z = 1]P(v = v). \qquad (2.16)$$

Assumption MMAR is non-refutable. $\qquad\qquad\qquad\qquad\qquad\qquad\square$

Proof: The Law of Iterated Expectations gives

$$E[g(y)] = \sum_{v \in V} E[g(y)|v = v]P(v = v). \tag{2.17}$$

Assumption MMAR states that

$$E[g(y)|v] = E[g(y)|v, z = 1]. \tag{2.18}$$

Applying (2.18) to (2.17) yields (2.16). The right side of (2.16) is point-identified by the sampling process, so $E[g(y)]$ is point-identified. Assumption MMAR is non-refutable because the empirical evidence reveals nothing about $E[g(y)|v, z = 0]$.

Q. E. D.

Proposition 2.4: (a) Let assumption MI hold. Then the closed interval

$$H_{MI}\{E[g(y)]\} =$$

$$[\max_{v \in V} E[g(y)z + g_0(1 - z)|v = v], \ \min_{v \in V} E[g(y)z + g_1(1 - z)|v = v]].$$

$$\tag{2.19}$$

is the identification region for $E[g(y)]$.

(b) Let $H_{MI}\{E[g(y)]\}$ be empty. Then assumption MI does not hold. □

Proof: (a) Application of equation (1.9′) to each conditional expectation $E[g(y)|v = v]$, $v \in V$ gives its identification region using the empirical evidence alone; that is, the closed interval

$$H\{E[g(y)|v = v]\} = [E[g(y)z + g_0(1 - z)|v = v], E[g(y)z + g_1(1 - z)|v = v]].$$

$$\tag{2.20}$$

Moreover, the identification region for $\{E[g(y)|v = v], v \in V\}$ is the $|V|$-dimensional rectangle $\times_{v \in V} H\{E[g(y)|v = v]\}$.

Assumption MI states that the expectations $E[g(y)|v = v]$, $v \in V$ coincide, all being equal to $E[g(y)]$. Hence $E[g(y)]$ must lie in $\cap_{v \in V} H\{E[g(y)|v = v]\}$. Any value in this set is feasible, so $H_{MI}\{E[g(y)]\}$ is the identification region.

(b) If assumption MI holds, the set $H_{MI}\{E[g(y)]\}$ is necessarily non-empty. Hence, the assumption cannot hold if $H_{MI}\{E[g(y)]\}$ is empty.

Q. E. D.

As with assumption SI, the identifying power of assumption MI can range from point identification to no power at all, depending on the nature of the instrumental variable. Point identification of $E[g(y)]$ occurs if there exists a $v \in V$ such that $P(z = 1 | v = v)$; then $E[g(y)] = E[g(y) | v = v]$. There is no identifying power if the pair (y, z) is statistically independent of v. Then $H_{MI}\{E[g(y)]\} = H\{E[g(y)]\}$.

Mean Monotonicity

Although mean independence is a weaker property than statistical independence, empirical researchers often find that assertions of mean independence are still too strong to be credible. There is therefore reason to ask whether Assumptions MMAR and MI may be weakened in ways that enhance credibility while preserving some identifying power. A simple way to do this is to change the equalities in equations (2.4) and (2.5) to the weak inequalities in equations (2.6) and (2.7).

Weakening assumption MMAR in this way yields assumption MMM, which asserts that, for each realization of v, the mean value of $g(y)$ when y is observed is greater than or equal to the mean value of $g(y)$ when y is missing. (The direction of the inequality can be reversed by applying the assumption to the function $-g(y)$.) Weakening assumption MI in this way yields assumption MM, which presumes that the set V has been pre-ordered. Propositions 2.5 and 2.6 characterize the identifying power of these monotonicity assumptions.

Proposition 2.5: Let assumption MMM hold. Then the identification region for $E[g(y)]$ is the closed interval

$$H_{MMM}\{E[g(y)]\} =$$

$$[E[g(y)|z=1]P(z=1) + g_0 P(z=0), \sum_{v \in V} E[g(y)|v=v, z=1]P(v=v)].$$

$$(2.21)$$

Assumption MMM is non-refutable.   □

Proof: Let $v \in V$. Under assumption MMM, the identification region for $E[g(y)|v=v, z=0]$ is the closed interval

$$H_{MMM}\{E[g(y)|v=v, z=0]\} = [g_0, E[g(y)|v=v, z=1]].   (2.22)$$

Moreover, the joint identification region for $\{E[g(y)|v=v, z=0], v \in V\}$

is the $|V|$-dimensional rectangle $\times_{v \in V} H_{MMM}\{E[g(y)|v = v, z = 0]\}$. The Law of Iterated Expectations gives

$$E[g(y)] = \sum_{v \in V} E[g(y)|v = v, z = 1]P(v = v, z = 1)$$

$$+ E[g(y)|v = v, z = 0]P(v = v, z = 0). \quad (2.23)$$

Applying (2.22) to (2.23) yields (2.21). Assumption MMM is non-refutable because the empirical evidence reveals nothing about $\{E[g(y)|v = v, z = 0], v \in V\}$.

Q. E. D.

Proposition 2.6: (a) Let V be an ordered set. Let assumption MM hold. Then the identification region for $E[g(y)]$ is the closed interval

$$H_{MM}\{E[g(y)]\} = [\sum_{v \in V} P(v = v)\{\max_{v' \leq v} E[g(y)z + g_0(1 - z)|v = v'\}],$$

$$\sum_{v \in V} P(v = v) \{\min_{v' \geq v} E[g(y)z + g_1(1 - z)|v = v'\}].$$

$$(2.24)$$

(b) Let $H_{MM}\{E[g(y)]\}$ be empty. Then assumption MM does not hold. □

Proof: (a) The proof to Proposition 2.4 showed that, using the empirical evidence alone, the identification region for the expectations $\{E[g(y)|v = v], v \in V\}$ is the $|V|$-dimensional rectangle $\times_{v \in V} H\{E[g(y)|v = v]\}$. Under assumption MM, a point $d \in R^V$ belongs to the identification region for $\{E[g(y)|v = v], v \in V\}$ if and only if d is an element of this rectangle whose components $(d_1, d_2, \ldots, d_{|V|})$ form a weakly increasing sequence. Applying this to the Law of Iterated Expectations (2.17) yields (2.24).

(b) If assumption MM holds, the set $H_{MM}\{E[g(y)]\}$ is necessarily non-empty. Hence, the assumption cannot hold if $H_{MM}\{E[g(y)]\}$ is empty.

Q. E. D.

Proposition 2.5 shows that, under assumption MMM, the identification region for $E[g(y)]$ is a right-truncated subset of the region obtained using the empirical evidence alone. The smallest feasible value of $E[g(y)]$ is the same as when using the empirical evidence alone. The largest is the value that $E[g(y)]$ would take under assumption MMAR.

Proposition 2.6 shows that the identification region under assumption MM is a subset of the region obtained using the empirical evidence alone and a superset of the one obtained under assumption MI. The identifying power of assumption MM depends on how the regions $\{H\{E[g(y)|v=v]\}$, $v \in V\}$ vary with v. The extreme possibilities occur if this sequence of intervals shifts to the left or right as v increases. In the former case, the identification region under assumption MM is the same as under assumption MI. In the latter case, assumption MM has no identifying power.

2.6. Other Assumptions Using Instrumental Variables

This chapter has examined assumptions that help to identify outcome distributions when an instrumental variable is observed. The tension between the credibility and strength of conclusions is especially evident as one weakens assumption SI to assumption MI and then to assumption MM. Each successive assumption is more plausible but has less identifying power.

It is easy to think of other assumptions that make different tradeoffs between credibility and identifying power. For example, assumption MI could be weakened not to the monotonicity asserted in assumption MM but rather to some form of "approximate" mean independence. A way to formalize this is to assert that, for all pairs $(v, v') \in V \times V$,

$$\|E[g(y)|v=v'] - E[g(y)|v=v]\| \leq C, \tag{2.25}$$

where $C > 0$ is a specified constant. Recall that the empirical evidence alone restricts the vector of expectations $E[g(y)|v]$ to the $|V|$-dimensional rectangle $\times_{v \in V} H\{E[g(y)|v=v]\}$. Relationship (2.25) further restricts $E[g(y)|v]$ to points in $R^{|V|}$ that satisfy specified linear inequalities.

Alternatively, assumption MI could be weakened to the zero-covariance assumption

$$E[g(y) \cdot v] - E[g(y)]E(v) = 0, \tag{2.26}$$

which may be rewritten as

$$\sum_{v \in V} P(v=v) [v - E(v)]E[g(y)|v=v] = 0. \tag{2.27}$$

The empirical evidence point-identifies $E(v)$. Hence, equation (2.27) establishes a linear constraint among the elements of $E[g(y)|v]$.

Complement 2A. Estimation with Nonresponse Weights

Organizations conducting major surveys commonly release public data files that provide *nonresponse weights* to be used for estimating means and other parameters of outcome distributions when data are missing. Nonresponse weights are distinct from *design weights*, which are used to compensate for planned variation in sampling rates across strata of the population.

The standard construction of nonresponse weights presumes the existence of an instrumental variable v. The standard use of such weights to infer a population mean $E[g(y)]$ yields a consistent estimate if assumption MMAR holds but not otherwise. Hence, empirical researchers contemplating application of nonresponse weights need to take care.

Weighted Sample Averages

Suppose that a random sample of size N has been drawn from population J. Let N(1) denote the sample members for whom $z = 1$, and let N_1 be the cardinality of N(1). Let $s(v): V \to [0, \infty)$ be a weighting function. Consider estimation of $E[g(y)]$ by the weighted sample average

$$\theta_N \equiv \frac{1}{N_1} \sum_{i \in N(1)} s(v_i) \cdot g(y_i).$$

By the Strong Law of Large Numbers, $\lim_{N \to \infty} \theta_N =_{a.s.} E[s(v) \cdot g(y) | z = 1]$. The standard weights provided by survey organizations have the form

$$s(v) = \frac{P(v = v)}{P(v = v | z = 1)}, \qquad v \in V.$$

With such weights,

$$E[s(v) \cdot g(y) | z = 1] = \sum_{v \in V} E[s(v) \cdot g(y) | v = v, z = 1] \cdot P(v = v | z = 1)$$

$$= \sum_{v \in V} E[g(y) | v = v, z = 1] \cdot P(v = v)$$

$$= \sum_{v \in V} E[g(y) | v = v, z = 1] \cdot P(v = v | z = 1) \cdot P(z = 1)$$

$$+ \sum_{v \in V} E[g(y) | v = v, z = 1] \cdot P(v = v | z = 0) \cdot P(z = 0)$$

$$= E[g(y) \,|\, z = 1] \cdot P(z = 1)$$

$$+ \sum_{v \in V} E[g(y) \,|\, v = v, z = 1] \cdot P(v = v \,|\, z = 0) \cdot P(z = 0).$$

The right side of this equation equals $E[g(y)]$ if assumption MMAR holds, but it generically differs from $E[g(y)]$ otherwise.

Endnotes

Sources and Historical Notes

My work on the identifying power of assumptions using instrumental variables began with Proposition 2.4, which was introduced in Manski (1990) and developed more fully in Manski (1994, Proposition 6). The monotonicity ideas in Propositions 2.5 and 2.6 are based on Manski and Pepper (2000, Proposition 1).

The term *instrumental variable* is due to Reiersol (1945) who, along with other econometricians of his time, studied the identification of linear structural equation systems. Goldberger (1972), in a review of this literature, dates the use of instrumental variables to identify linear structural equations back to Wright (1928). Modern econometric research uses instrumental variables to address this and many other identification problems. However, the practice invariably is to assert assumptions strong enough to yield point identification of quantities of interest.

It is revealing to consider some history within economics. Until the early 1970s, empirical researchers confronting missing outcome data essentially always used assumption (2.1), although often without explicit discussion. At that time, the credibility of this assumption was questioned sharply when researchers observed that, in many economic settings, the process by which observations on y become missing is related to the value of y; see, for example, Gronau (1974). Econometricians subsequently developed a variety of models of missing data that do not assert (2.1) but instead use instrumental variables and parametric restrictions on the shape of the distribution $P(y, z)$ to point-identify $P(y)$; see, for example, Heckman (1976) and Maddala (1983). These developments were initially greeted with widespread enthusiasm, but methodological studies soon showed that seemingly minor changes in the assumptions imposed could generate large changes in the implied value of $P(y)$; see, for example, Arabmazar and Schmidt (1982), Goldberger (1983), and Hurd (1979).

Text Notes

1. Proposition 2.1 has long been well-known, so much so that it is unclear when the idea originated. In the survey sampling literature, this proposition provides the basis for construction of sampling weights that aim to enable population inference in the presence of missing data; see Complement 2A. Rubin (1976) introduced the term *missing at random*. In applied econometric research, assumption (2.1) is sometimes called *selection on observables*; see Fitzgerald, Gottschalk, and Moffitt (1998, Section IIIA) for discussion of the history of the concept and term.

2. Empirical researchers sometimes assert that assumption MAR becomes more credible as the instrumental variable partitions the population into more refined sub-populations. That is, if v_1 and v_2 are alternative specifications of the instrumental variable, with $P(v_1|v_2)$ degenerate, researchers may assert that v_2 is a more credible instrumental variable than is v_1. Unfortunately, this assertion typically is backed up by nothing more than the empty statement that v_2 "controls for" more determinants of missing data than v_1. In principle, assumption MAR could hold for both, either, or neither of v_1 and v_2.

3

Conditional Prediction with Missing Data

3.1. Prediction of Outcomes Conditional on Covariates

A large part of statistical practice aims to predict outcomes conditional on covariates. Suppose that each member j of population J has an outcome y_j in a space Y and a covariate x_j in a space X. Let the random variable (y, x): $J \to Y \times X$ have distribution $P(y, x)$. In general terms, the objective is to learn the conditional distributions $P(y|x = x)$, $x \in X$. A particular objective may be to learn the conditional expectation $E(y|x = x)$, conditional median $M(y|x = x)$, or another point predictor of y conditional on an event $\{x = x\}$.

This chapter studies the feasibility of prediction when a sampling process draws persons at random from J and realizations of (y, x) may be observable in whole, in part, or not at all. Two binary random variables (z_y, z_x) now indicate observability. A realization of y is observable if $z_y = 1$ but not if $z_y = 0$; a realization of x is observable if $z_x = 1$ but not if $z_x = 0$. The sampling process reveals distributions $P(z_y, z_x)$, $P(y, x|z_y = 1, z_x = 1)$, $P(y|z_y = 1, z_x = 0)$, and $P(x|z_y = 0, z_x = 1)$. The problem is to use this empirical evidence to infer $P(y|x = x)$, $x \in X$.

In practice, empirical researchers may face complex patterns of missing data; some sample members may have missing outcome data, others may have missing covariate data, and others may have jointly missing outcomes and covariates. Nevertheless, it is instructive to study the polar cases in which all missing data are of the same type. Section 3.2 briefly reviews the case in which only outcome data are missing; here $P(z_x = 1) = 1$. Section 3.3 studies inference when all sample members with missing data have jointly missing outcomes and covariates; here $P(z_y = z_x = 1) + P(z_y = z_x = 0) = 1$. Section 3.4 supposes that only covariate data are missing, so $P(z_y = 1) = 1$.

40

With these polar cases understood, Section 3.5 examines general patterns of missing data. Throughout Sections 3.2 to 3.5, the object of interest is the distribution $P(y|x = x)$ evaluated at a given $x \in X$. Section 3.6 studies joint inference on the set of conditional distributions $[P(y|x = x), x \in X]$.

To simplify the presentation, I suppose throughout this chapter that realizations of outcomes or covariates are either completely observed or entirely missing. Thus, I do not consider interval measurement of (y, x) or partial observability of covariate vectors, with realizations of x having some components observed and others not. I also suppose that the covariate space X is finite and that $P(x = x, z_x = 1) > 0$ for all $x \in X$. These regularity conditions are maintained without further reference.

3.2. Missing Outcomes

Chapters 1 and 2 studied identification of the marginal distribution $P(y)$ when some realizations of y may be missing. The results obtained there apply immediately to $P(y|x = x)$ if realizations of x are always observable. One simply needs to redefine the population of interest to be the sub-population of J for which $\{x = x\}$. Then equation (1.2) shows that the identification region using the empirical evidence alone is

$H[P(y|x = x)] =$

$$[P(y|x = x, z_y = 1)P(z_y = 1|x = x) + \gamma P(z_y = 0|x = x), \gamma \in \Gamma_Y].$$

(3.1)

Similarly, the other findings reported in Chapters 1 and 2 continue to hold if one conditions all distributions on the event $\{x = x\}$ and replaces all occurrences of z with z_y.

3.3. Jointly Missing Outcomes and Covariates

Jointly missing outcomes and covariates is a regular occurrence in survey research. Realizations of (y, x) may be missing in their entirety when sample members refuse to be interviewed or cannot be contacted by survey administrators. Joint missingness also occurs when outcomes are missing and the objective is to learn a distribution of the form $P(y|y \in B)$, where B $\subset$ Y. If the outcome y is not observed, then the conditioning event $\{y \in B\}$ necessarily is not observed.

In principle, identification of $P(y|x = x)$ when (y, x) realizations are jointly missing may be studied as an instance of the problem of missing outcomes set out in Section 1.1. Let the outcome of interest be (y, x) rather than y alone. Let $\Gamma_{Y \times X}$ denote the space of all probability distributions on $Y \times X$. Let $z_{yx} = 1$ if $z_y = z_x = 1$ and $z_{yx} = 0$ otherwise. Then equation (1.2) yields the identification region for the joint distribution $P(y, x)$, namely

$$H[P(y, x)] = [P(y, x|z_{yx} = 1)P(z_{yx} = 1) + \gamma P(z_{yx} = 0), \gamma \in \Gamma_{Y \times X}]. \quad (3.2)$$

Now let $\tau(\cdot): \Gamma_{Y \times X} \to \Gamma_Y$ be the function that maps $P(y, x)$ into the distribution $P(y|x = x)$. Then equation (1.5) yields

$$H[P(y|x = x)] = \{\tau(\eta), \eta \in H[P(y, x)]\}. \quad (3.3)$$

Similarly, equation (1.6) may be applied if distributional assumptions have been imposed. Thus, all of the results obtained in Chapters 1 and 2 may, in principle, be used to study identification of $P(y|x = x)$ when (y, x) realizations are jointly missing.

Identification Using the Empirical Evidence Alone
Although equations (3.2) and (3.3) describe the identification region $H[P(y|x = x)]$ in principle, they do not provide a transparent description. Proposition 3.1 shows directly that the region has a simple structure.

Proposition 3.1: Let $P(z_y = z_x = 1) + P(z_y = z_x = 0) = 1$. Then

$$H[P(y|x = x)] = \{P(y|x = x, z_{yx} = 1)r(x) + \gamma[1 - r(x)], \gamma \in \Gamma_Y\}, \quad (3.4a)$$

where

$$r(x) \equiv \frac{P(x = x|z_{yx} = 1)P(z_{yx} = 1)}{P(x = x|z_{yx} = 1)P(z_{yx} = 1) + P(z_{yx} = 0)}. \quad (3.4b)$$

$\square$

Proof: The Law of Total Probability gives

$$P(y|x = x) =$$

$$P(y|x = x, z_{yx} = 1)P(z_{yx} = 1|x = x) + P(y|x = x, z_{yx} = 0)P(z_{yx} = 0|x = x).$$

$$(3.5)$$

For $i = 0$ or 1, Bayes Theorem gives

$$P(z_{yx} = i \,|\, x = x) = \frac{P(x = x \,|\, z_{yx} = i)P(z_{yx} = i)}{P(x = x \,|\, z_{yx} = 1)P(z_{yx} = 1) + P(x = x \,|\, z_{yx} = 0)P(z_{yx} = 0)}.$$

$$(3.6)$$

Inserting (3.6) into (3.5) yields

$$P(y \,|\, x = x) =$$

$$P(y \,|\, x = x, z_{yx} = 1) \frac{P(x = x \,|\, z_{yx} = 1)P(z_{yx} = 1)}{P(x = x \,|\, z_{yx} = 1)P(z_{yx} = 1) + P(x = x \,|\, z_{yx} = 0)P(z_{yx} = 0)}$$

$$+ \; P(y \,|\, x = x, z_{yx} = 0) \frac{P(x = x \,|\, z_{yx} = 0)P(z_{yx} = 0)}{P(x = x \,|\, z_{yx} = 1)P(z_{yx} = 1) + P(x = x \,|\, z_{yx} = 0)P(z_{yx} = 0)}.$$

$$(3.7)$$

Consider the right side of equation (3.7). The sampling process identifies $P(z_{yx})$, $P(x = x \,|\, z_{yx} = 1)$, and $P(y \,|\, x = x, z_{yx} = 1)$. It is uninformative about $P(x = x \,|\, z_{yx} = 0)$ and $P(y \,|\, x = x, z_{yx} = 0)$. Hence, the identification region for $P(y \,|\, x = x)$ is

$$H[P(y \,|\, x = x)] =$$

$$\bigcup_{p \in [0, 1]} \; \left\{ P(y \,|\, x = x, z_{yx} = 1) \; \frac{P(x = x \,|\, z_{yx} - 1)P(z_{yx} = 1)}{P(x = x \,|\, z_{yx} = 1)P(z_{yx} = 1) + pP(z_{yx} = 0)} \right.$$

$$+ \; \gamma \; \frac{pP(z_{yx} = 0)}{P(x = x \,|\, z_{yx} = 1)P(z_{yx} = 1) + pP(z_{yx} = 0)} \; , \; \left. \gamma \in \Gamma_Y \right\}.$$

$$(3.8)$$

For each $p \in [0, 1]$, the distributions in brackets are mixtures of $P(y \,|\, x = x, z_{yx} = 1)$ and arbitrary distributions on Y. The set of mixtures enlarges as p increases from 0 to 1. Hence, it suffices to set $p = 1$. This yields (3.4).

Q. E. D.

It is of interest to compare the identification region for $P(y|x = x)$ in (3.4) with the region (3.1) obtained when only realizations of y are missing. The two regions have the same form, with $r(x)$ here replacing $P(z = 1|x = x)$ there. The quantity $r(x)$ is the smallest feasible value of $P(z_{yx} = 1|x = x)$ and is obtained by conjecturing that all missing covariate realizations have the value x. Thus, joint missingness of (y, x) exacerbates the identification problem produced by missingness of y alone.

The degree to which joint missingness exacerbates the identification problem depends on the prevalence of the value x among the observable realizations of x. Inspection of (3.4b) shows that $r(x) = P(z_{yx} = 1)$ when $P(x = x|z_{yx} = 1) = 1$ and decreases to zero as $P(x = x|z_{yx} = 1)$ falls to zero. Thus, region (3.4) is uninformative if the observable covariate distribution $P(x|z_{yx} = 1)$ places zero mass on the value x.

With $r(x)$ replacing $P(z = 1|x = x)$ and z_{yx} replacing z, all results obtained in Chapter 1 hold when realizations of (y, x) are jointly missing. Proposition 1.2 has this analog.

Proposition 3.2: Let D respect stochastic dominance. Let $g \in G$. Let $P(z_y = z_x = 1) + P(z_y = z_x = 0) = 1$. Then the smallest and largest points in the identification region for $D\{P[g(y)]\}$ are $D\{P[g(y)|z_{yx} = 1]r(x) + \gamma_{0g}[1 - r(x)]\}$ and $D\{P[g(y)|z_{yx} = 1]r(x) + \gamma_{1g}[1 - r(x)]\}$.   □

Proposition 1.3 has the following generalization to settings in which data are available from multiple sampling processes, with only outcomes missing in some sampling processes and (y, x) jointly missing in the others.

Proposition 3.3: Let there be a set M of sampling processes for which $P(z_x = 1) = 1$ and a set M' of sampling processes for which $P(z_y = z_x = 1) + P(z_y = z_x = 0) = 1$. Then

$$H_{(M, M')}[P(y|x = x)] =$$

$$\bigcap_{m \in M} [P(y|x = x, z_{my} = 1)P(z_{my} = 1|x = x) + \gamma_m P(z_{my} = 0|x = x), \; \gamma_m \in \Gamma_Y]$$

$$\bigcap_{m \in M'} [P(y|x = x, z_{myx} = 1)r_m(x) + \gamma_m [1 - r_m(x)], \; \gamma_m \in \Gamma_Y]$$

$$(3.9)$$

is the identification region for $P(y|x = x)$.   □

Distributional Assumptions

When (y, x) are jointly missing, empirical researchers often assume that observed and missing outcomes have the same distribution conditional on the event $\{x = x\}$; that is,

$$P(y|x = x) = P(y|x = x, z_{yx} = 0) = P(y|x = x, z_{yx} = 1). \quad (3.10)$$

Suppose that $P(z_{yx} = 1) > 0$. Then $P(y|x = x, z_{yx} = 1)$ is revealed by the sampling process, so $P(y|x=x)$ is point-identified under assumption (3.10). However, the credibility of this non-refutable assumption is often suspect.

Distributional assumptions that use instrumental variables have identifying power when (y, x) are jointly missing. Consider Assumptions MAR and SI. Applied to the sub-population of J for which $\{x = x\}$, these assumptions are respectively

$$P(y|v, x = x) = P(y|v, x = x, z_{yx} = 0) = P(y|v, x = x, z_{yx} = 1) \quad (3.11)$$

and

$$P(y|v, x = x) = P(y|x = x). \quad (3.12)$$

The identifying power of assumption SI follows easily from Propositions 3.1 and 2.2. Let $v \in V$. Applying Proposition 3.1 to the sub-population of J with $\{v = v, x = x\}$ yields

$$H[P(y|v = v, x = x)] =$$

$$\{P(y|v = v, x = x, z_{yx} = 1)r(v, x) + \gamma_v[1 - r(v, x)], \quad \gamma_v \in \Gamma_Y\},$$

$$(3.13a)$$

where

$$r(v, x) \equiv$$

$$\frac{P(x = x|v = v, z_{yx} = 1)P(z_{yx} = 1|v = v)}{P(x = x|v = v, z_{yx} = 1)P(z_{yx} = 1|v = v) + P(z_{yx} = 0|v = v)}.$$

$$(3.13b)$$

Emulation of the proof to Proposition 2.2 then yields the following proposition.

Proposition 3.4: (a) Let assumption SI hold, as in (3.12). Let $P(z_y = z_x = 1)$ $+ P(z_y = z_x = 0) = 1$. Then the identification region for $P(y|x = x)$ is

$$H_{SI}[P(y|x = x)] =$$

$$\bigcap_{v \in V} \{P(y|v = v, x = x, z_{yx} = 1)r(v, x) + \gamma_v[1 - r(v, x)], \; \gamma_v \in \Gamma_Y\}.$$

$$(3.14)$$

(b) Let $H_{SI}[P(y| x = x)]$ be empty. Then (3.12) does not hold.   □

It is a more complex matter to determine the identifying power of assumption MAR when (y, x) are jointly missing. When (3.11) holds, emulation of the proof to Proposition 2.1 shows that

$$P(y|x = x) = \sum_{v \in V} P(y|v = v, x = x, z_{yx} = 1)P(v = v|x = x). \quad (3.15)$$

If only outcomes were missing, the empirical evidence would reveal all quantities on the right side of (3.15). However, with (y, x) jointly missing, the empirical evidence does not reveal $P(v|x = x)$; hence $P(y|x = x)$ is not point-identified.

To describe the identification region for $P(y|x = x)$, we need to know the identification region for $P(v|x = x)$. Given that v is always observed, inference on $P(v|x = x)$ is a problem of prediction when only covariates are missing. This problem is the subject of the next section.

3.4. Missing Covariates

Suppose now that realizations of the outcome y are always observed but realizations of the covariate x may be missing. Proposition 3.5 gives the identification region for $P(y|x = x)$ using the empirical evidence alone.

Proposition 3.5: Let $P(z_y = 1) = 1$. Then

$$H[P(y|x = x)] = \bigcup_{p \in [0, 1]}$$

$$\left\{ P(y|x = x, z_x = 1) \; \frac{P(x = x|z_x = 1)P(z_x = 1)}{P(x = x|z_x = 1)P(z_x = 1) + pP(z_x = 0)} \right.$$

$$+ \eta \ \frac{pP(z_x = 0)}{P(x = x \,|\, z_x = 1)P(z_x = 1) + pP(z_x = 0)} \ , \ \eta \in \Gamma_Y(p) \},$$

$$(3.16a)$$

where

$$\Gamma_Y(p) \equiv \Gamma_Y \cap \{[P(y \,|\, z_x = 0) - \gamma(1 - p)]/p, \ \gamma \in \Gamma_Y\}. \qquad (3.16b)$$

$\square$

Proof: Applying the Law of Total Probability and Bayes Theorem as in (3.5) and (3.6) yields

$$P(y \,|\, x = x) \ =$$

$$P(y \,|\, x = x, z_x = 1) \ \frac{P(x = x \,|\, z_x = 1)P(z_x = 1)}{P(x = x \,|\, z_x = 1)P(z_x = 1) + P(x = x \,|\, z_x = 0)P(z_x = 0)}$$

$$+ \ P(y \,|\, x = x, z_x = 0) \ \frac{P(x = x \,|\, z_x = 0)P(z_x = 0)}{P(x = x \,|\, z_x = 1)P(z_x = 1) + P(x = x \,|\, z_x = 0)P(z_x = 0)} \ .$$

$$(3.17)$$

Of the quantities on the right side of equation (3.17), the sampling process reveals $P(z_x)$, $P(x = x \,|\, z_x = 1)$, and $P(y \,|\, x = x, z_x = 1)$, but not $P(x = x \,|\, z_x = 0)$ and $P(y \,|\, x = x, z_x = 0)$. It does reveal $P(y \,|\, z_x = 0)$, which is related to $P(x = x \,|\, z_x = 0)$ and $P(y \,|\, x = x, z_x = 0)$ by the Law of Total Probability

$$P(y \,|\, z_x = 0) \ =$$

$$P(y \,|\, x = x, z_x = 0)P(x = x \,|\, z_x = 0) \ + P(y \,|\, x \neq x, z_x = 0)P(x \neq x \,|\, z_x = 0).$$

$$(3.18)$$

To determine the identifying power of (3.18), suppose that $P(x = x \,|\, z_x = 0)$ = p. Then $\Gamma_Y(p)$ given in (3.16b) is the set of values of $P(y \,|\, x = x, z_x = 0)$ that are consistent with (3.18). Now let p range over the interval [0, 1]. This yields (3.16a).

Q. E. D.

The restriction imposed by equation (3.18) makes the problem of missing covariates qualitatively different from the missing data problems studied earlier in this chapter.[1] For given $p \in [0, 1]$, the set of distributions $\Gamma_Y(p)$ are the solutions to a mixture problem that will be studied in depth in Chapter 4. I defer discussion of the structure of $\Gamma_Y(p)$ until then.

Missing covariates pose a less severe observational problem than do jointly missing outcomes and covariates. Hence, the identification region derived in Proposition 3.5 necessarily is a subset of the one obtained in Proposition 3.1. Comparison of (3.16) with (3.8) makes this precise. Whereas γ could range over the space Γ_Y of all distributions in (3.8), it can only range over the restricted spaces $\Gamma_Y(p)$, $p \in [0, 1]$ in (3.16).

Indeed, $P(y|x = x)$ may even be point-identified. This happens if $P(y|x = x, z_x = 1) = P(y|z_x = 0)$, and these distributions are degenerate. Then region (3.16) contains only one element, $P(y|x = x, z_x = 1)$.

Assumptions MAR and SI

Proposition 3.5 enables description of the identifying power of assumption MAR when (y, x) are jointly missing, a question that was left open in Section 3.3. Recall equation (3.15). Assuming that $P(z_{yx} = 1) > 0$, the only quantity not identified on the right side of (3.15) is $P(v|x = x)$. Hence, joint missingness of (y, x) has the same consequences as missingness of x alone. Proposition 3.5 gives the identification region for $P(v|x = x)$. From this, Proposition 3.6 follows.

Proposition 3.6: Let assumption MAR hold, as in (3.11). Let $P(z_{yx} = 1) > 0$. Let $H[P(v|x = x)]$ be the identification region applying Proposition 3.5 to $P(v|x = x)$. Then the identification region for $P(y|x = x)$ is

$$H_{MAR}[P(y|x = x)] =$$

$$\{ \textstyle\sum_{v \in V} P(y|v = v, x = x, z_{yx} = 1)\eta(v = v), \ \eta \in H[P(v|x = x)]\}. \quad (3.19)$$

$\square$

Proposition 3.5 immediately gives the identifying power of assumption SI when covariates are missing. This result is Proposition 3.7.

Proposition 3.7: (a) Let assumption SI hold, as in (3.12). Let $P(z_y = 1) = 1$. For $v \in V$, let $H[P(y|v = v, x = x)]$ be the identification region obtained by applying Proposition 3.5 to $P(y|v = v, x = x)$. Then the identification region for $P(y|x = x)$ is

$$H_{SI}[P(y\,|\,x = x)] \;=\; \bigcap_{v \in V} H[P(y\,|\,v = v, x = x)]. \qquad (3.20)$$

(b) Let $H_{SI}[P(y\,|\,x = x)]$ be empty. Then (3.12) does not hold.   □

3.5. General Missing-Data Patterns

Consider now a sampling process with a general pattern of missing data in which some realizations of (y, x) may be completely observed, others observed in part, and still others not observed at all. The structure of the problem of inference on $P(y\,|\,x = x)$ is displayed by the Law of Total Probability and Bayes Theorem, which give

$$P(y\,|\,x = x) \;=$$

$$\sum_j \sum_k P(y\,|\,x = x, z_x = j, z_y = k) \; \frac{P(x = x\,|\,z_x = j, z_y = k)P(z_x = j, z_y = k)}{\sum_\ell \sum_m P(x = x\,|\,z_x = \ell, z_y = m)P(z_x = \ell, z_y = m)}.$$

$$(3.21)$$

Examine the right side of (3.21). The sampling process identifies $P(z_x, z_y)$, $P(x = x\,|\,z_x = 1, z_y)$, and $P(y\,|\,x = x, z_x = 1, z_y = 1)$. It does not identify $P(x = x\,|\,z_x = 0, z_y)$, $P(y\,|\,x = x, z_x = 0, z_y = 1)$, or $P(y\,|\,x = x, z_x, z_y = 0)$. The sampling process does, however, reveal $P(y\,|\,z_x = 0, z_y = 1)$, which is related to $P(x = x\,|\,z_x = 0, z_y = 1)$ and $P(y\,|\,x = x, z_x = 0, z_y = 1)$ by the Law of Total Probability

$$P(y\,|\,z_x = 0, z_y = 1) \;=\; P(y\,|\,x = x, z_x = 0, z_y = 1)P(x = x\,|\,z_x = 0, z_y = 1)$$

$$+ \; P(y\,|\,x \neq x, z_x = 0, z_y = 1)P(x \neq x\,|\,z_x = 0, z_y = 1).$$

$$(3.22)$$

Thus, using the empirical evidence alone, the identification region for $P(y\,|\,x = x)$ has the following form.

Proposition 3.8: Let $P_{jk} \equiv P(z_x = j, z_y = k)$ for $j, k = 0$ or 1. Then

$$H[P(y\,|\,x = x)] \;=$$

$$\{ P(y \,|\, x = x, z_x = 1, z_y = 1) \; \frac{P(x = x \,|\, z_x = 1, z_y = 1)P_{11}}{\sum_k P(x = x \,|\, z_x = 1, z_y = k)P_{1k} + p_0 P_{00} + p_1 P_{01}}$$

$$+ \quad \eta_{10} \; \frac{P(x = x \,|\, z_x = 1, z_y = 0)P_{10}}{\sum_k P(x = x \,|\, z_x = 1, z_y = k)P_{1k} + p_0 P_{00} + p_1 P_{01}}$$

$$+ \quad \eta_{00} \; \frac{p_0 P_{00}}{\sum_k P(x = x \,|\, z_x = 1, z_y = k)P_{1k} + p_0 P_{00} + p_1 P_{01}}$$

$$+ \quad \eta_{01} \; \frac{p_1 P_{01}}{\sum_k P(x = x \,|\, z_x = 1, z_y = k)P_{1k} + p_0 P_{00} + p_1 P_{01}},$$

$$(\eta_{10}, \eta_{00}, \eta_{01}) \in \Gamma_Y \times \Gamma_Y \times \Gamma_Y(p_1) ; \quad (p_0, p_1) \in [0, 1]^2 \},$$

$$(3.23a)$$

where

$$\Gamma_Y(p_1) \equiv \Gamma_Y \cap \{[P(y \,|\, z_x = 0, z_y = 1) - \gamma(1 - p_1)]/p_1, \; \gamma \in \Gamma_Y\}. \qquad (3.23b)$$

□

This identification region is generally complex whenever there is positive probability P_{01} that realizations of outcomes are observable but those of covariates are not. Nevertheless, Proposition 3.8 yields a relatively simple closed-form identification region for the probability $P(y \in B \,|\, x = x)$ that y lies in any set B. Corollary 3.8.1 gives this result.[2]

Corollary 3.8.1: Let B be a non-empty, proper, and measurable subset of Y. Define

$$R(x) \equiv P(y \in B \,|\, x = x, z_x = 1, z_y = 1)P(x = x \,|\, z_x = 1, z_y = 1)P_{11}$$

$$+ P(x = x \,|\, z_x = 1, z_y = 0)P_{10} + P_{00} + P(y \in B \,|\, z_x = 0, z_y = 1)P_{01},$$

$$S(x) \equiv \sum_k P(x = x \mid z_x = 1, z_y = k)P_{1k} + P_{00} + P(y \in B \mid z_x = 0, z_y = 1)P_{01},$$

$$T(x) \equiv \sum_k P(x = x \mid z_x = 1, z_y = k)P_{1k} + P_{00} + P(y \notin B \mid z_x = 0, z_y = 1)P_{01},$$

$$L(x) \equiv \frac{P(y \in B \mid x = x, z_x = 1, z_y = 1)P(x = x \mid z_x = 1, z_y = 1)P_{11}}{T(x)},$$

and $U(x) \equiv R(x)/S(x)$. Then the identification region for $P(y \in B \mid x = x)$ is

$$H[P(y \in B \mid x = x)] = [L(x), U(x)]. \tag{3.24}$$

$\square$

Proof: Proposition 3.8 shows that

$$H[P(y \in B \mid x = x)] =$$

$$\left\{ P(y \in B \mid x = x, z_x = 1, z_y = 1) \frac{P(x = x \mid z_x = 1, z_y = 1)P_{11}}{\sum_k P(x = x \mid z_x = 1, z_y = k)P_{1k} + p_0 P_{00} + p_1 P_{01}} \right.$$

$$+ \quad \eta_{10}(B) \frac{P(x = x \mid z_x = 1, z_y = 0)P_{10}}{\sum_k P(x = x \mid z_x = 1, z_y = k)P_{1k} + p_0 P_{00} + p_1 P_{01}}$$

$$+ \quad \eta_{00}(B) \frac{p_0 P_{00}}{\sum_k P(x = x \mid z_x = 1, z_y = k)P_{1k} + p_0 P_{00} + p_1 P_{01}}$$

$$+ \quad \eta_{01}(B) \frac{p_1 P_{01}}{\sum_k P(x = x \mid z_x = 1, z_y = k)P_{1k} + p_0 P_{00} + p_1 P_{01}},$$

$$\left[\eta_{10}(B), \eta_{00}(B), \eta_{01}(B)\right] \in [0, 1]^2 \times I(p_1); \ (p_0, p_1) \in [0, 1]^2 \left. \right\},$$

$$\tag{3.25a}$$

where

$$I(p_1) \equiv [0, 1] \cap \{[P(y \in B \mid z_x = 0, z_y = 1) - \lambda(1-p_1)]/p_1, \lambda \in [0, 1]\}. \quad (3.25b)$$

Hold p_1 fixed, and vary λ over its range $[0, 1]$. This shows that

$$\eta_{01}(B) \in [\max\{0, [A - (1 - p_1)]/p_1\}, \min\{1, A/p_1\}], \quad (3.26)$$

where $A \equiv P(y \in B \mid z_x = 0, z_y = 1)$. Continue to hold p_1 fixed and vary $[\eta_{10}(B), \eta_{00}(B), p_0] \in [0, 1]^3$. This shows that

$$P(y \in B \mid x = x) \in [L^*(x, p_1), U^*(x, p_1)], \quad (3.27a)$$

where

$$L^*(x, p_1) \equiv$$

$$P(y \in B \mid x = x, z_x = 1, z_y = 1) \; \frac{P(x = x \mid z_x = 1, z_y = 1)P_{11}}{\sum_k P(x = x \mid z_x = 1, z_y = k)P_{1k} + P_{00} + p_1 P_{01}}$$

$$+ \; \max\{0, [A - (1 - p_1)]/p_1\} \; \frac{p_1 P_{01}}{\sum_k P(x = x \mid z_x = 1, z_y = k)P_{1k} + P_{00} + p_1 P_{01}}$$

$$(3.27b)$$

and

$$U^*(x, p_1) \equiv$$

$$P(y \in B \mid x = x, z_x = 1, z_y = 1) \; \frac{P(x = x \mid z_x = 1, z_y = 1)P_{11}}{\sum_k P(x = x \mid z_x = 1, z_y = k)P_{1k} + P_{00} + p_1 P_{01}}$$

$$+ \; \frac{P(x = x \mid z_x = 1, z_y = 0)P_{10} + P_{00}}{\sum_k P(x = x \mid z_x = 1, z_y = k)P_{1k} + P_{00} + p_1 P_{01}}$$

$$+ \quad \min(1, A/p_1) \quad \frac{p_1 P_{01}}{\sum\limits_{k} P(x = x \mid z_x = 1, z_y = k) P_{1k} + P_{00} + p_1 P_{01}} \; .$$

$$(3.27c)$$

Finally, minimize $L^*(x, p_1)$ and maximize $U^*(x, p_1)$ over $p_1 \in [0, 1]$. The function $L^*(x, \cdot)$ is unimodal with unique minimum at $p_1 = 1 - A$. This yields the overall lower bound $L(x) = L^*(x, 1 - A)$. The function $U^*(x, \cdot)$ is unimodal with unique maximum at $p_1 = A$. This yields the overall upper bound $U(x) = U^*(x, A)$.

Q. E. D.

3.6. Joint Inference on Conditional Distributions

In Sections 3.2 through 3.5, the objective was presumed to be prediction of the outcome y conditional on the covariate x taking a specified value x; hence the object of interest was the conditional distribution $P(y \mid x = x)$. Researchers often want to predict outcomes when covariates take multiple values. Then the object of interest is the set of conditional distributions $[P(y \mid x = x), x \in X]$ or some functional thereof.

The identification region for $[P(y \mid x = x), x \in X]$ necessarily is a subset of the Cartesian product of the identification regions for each component distribution. Using the empirical evidence alone, that is

$$H[P(y \mid x = x), x \in X] \subset \times_{x \in X} H[P(y \mid x = x)]. \qquad (3.28)$$

Relationship (3.28) follows immediately from the definition of an identification region. Region $H[P(y \mid x = x), x \in X]$ gives all jointly feasible values of $[P(y \mid x = x), x \in X]$. For each $x \in X$, $H[P(y \mid x = x)]$ gives all feasible values of $P(y \mid x = x)$. Joint feasibility implies component-by-component feasibility, so (3.28) must hold.

To go beyond (3.28) in characterizing the problem of joint inference, one must specify the nature of the missing-data problem. The structure of the joint identification region is complex for sampling processes with general patterns of missing data, but simple results hold if only outcomes are missing or if (y, x) are jointly missing.[3]

Missing Outcomes

Suppose that only outcome data are missing and that no distributional assumptions are imposed. The Law of Total Probability gives

$$[P(y|x = x), x \in X] =$$

$$[P(y|x = x, z_y = 1)P(z_y = 1|x = x) + P(y|x = x, z_y = 0)P(z_y = 0|x = x), \ x \in X].$$

$$(3.29)$$

The sampling process identifies all quantities on the right side of (3.29) except for the set of distributions $[P(y|x = x, z_y = 0), x \in X]$, which can take any value in $\times_{x \in X} \Gamma_Y$. Hence, we have Proposition 3.9.

Proposition 3.9: Let $P(z_x = 1) > 0$. Then

$$H[P(y|x = x), x \in X] = \times_{x \in X} H[P(y|x = x)]$$

$$= \times_{x \in X} [P(y|x = x, z_y = 1)P(z_y = 1|x = x) + \gamma_x P(z_y = 0|x = x), \gamma_x \in \Gamma_Y].$$

$$(3.30)$$
□

Analogous findings hold if data from multiple sampling processes are available or if distributional assumptions using instrumental variables are imposed. In all of the settings considered in Chapters 1 and 2, joint inference on $[P(y|x = x), x \in X]$ is equivalent to component-by-component inference on $P(y|x = x), x \in X$.

Jointly Missing Outcomes and Covariates

When realizations of (y, x) are jointly missing, inference on the set of distributions $[P(y|x = x), x \in X]$ is not equivalent to component-by-component inference on $P(y|x = x), x \in X$. The reason is that a missing covariate realization cannot simultaneously have multiple values.

Recall Proposition 3.1, which gave the identification region for $P(y|x = x)$ at a specified value x. Equation (3.8) showed that the set of feasible values for $P(y|x = x)$ enlarges with the probability $P(x = x|z_{yx} = 0)$ that a missing realization of x takes the value x; hence $H[P(y|x = x)]$ emerged by setting $P(x = x|z_{yx} = 0) = 1$.

Now consider any other covariate value x'. If $P(x = x|z_{yx} = 0) = 1$, then $P(x = x'|z_{yx} = 0) = 0$. Hence, by (3.7), $P(y|x = x') = P(y|x = x', z_{yx} = 1)$. Thus $P(y|x = x)$ can range over its entire identification region only if the distributions $[P(y|x = x'), x' \in X, x' \neq x]$ take particular values. Thus

$H[P(y|x = x), x \in X]$ is a proper subset of $\times_{x \in X} H[P(y|x = x)]$.
   Proposition 3.10 characterizes the joint region.

Proposition 3.10: Let $P(z_y = z_x = 1) + P(z_y = z_x = 0) = 1$. Let S denote the unit simplex in $R^{|X|}$. Then

$$H[P(y|x = x), x \in X] = \bigcup_{(p_x, x \in X) \in S}$$

$$\{\times_{x \in X} \ [P(y|x = x, z_{yx} = 1) \ \frac{P(x = x|z_{yx} = 1)P(z_{yx} = 1)}{P(x = x|z_{yx} = 1)P(z_{yx} = 1) + p_x P(z_{yx} = 0)}$$

$$+ \ \gamma_x \ \frac{p_x \ P(z_{yx} = 0)}{P(x = x|z_{yx} = 1)P(z_{yx} = 1) + p_x \ P(z_{yx} = 0)} , \ \gamma_x \in \Gamma_Y]\}. \quad (3.31)$$

$\square$

Proof: The vector $[P(x = x|z_{yx} = 0), x \in X]$ can take any value in the unit simplex. For any feasible value of this vector, the set of feasible values of $[P(y|x = x), x \in X]$ is the Cartesian product of the sets of distributions in brackets in (3.31).

Q. E. D.

Complement 3A. Unemployment Rates

Complement 1A used NLSY data to estimate the probability that a member of the surveyed population was employed in 1991. Now consider the problem of inference on the official unemployment rate as measured in the United States by the Bureau of Labor Statistics. This rate is the probability of unemployment within the sub-population of persons who are in the labor force. When the 1991 employment status of an NLSY sample member is not reported, data are missing not only on that person's unemployment outcome but also on his or her membership in the labor force. Thus, inference on the official unemployment rate poses a problem of jointly missing outcome and covariate data.

   As in Complement 1A, the quantity of interest is $P[y = 1|y \in \{1, 2\}]$ or, perhaps, $P[y = 1|BASE, y \in \{1, 2\}]$. The data in Table 1.1 show that the empirical unemployment rate among the individuals who responded to the

1991 employment-status question and who reported that they were in the labor force is $P[y = 1 | y \in \{1, 2\}, z = 1] = 297/4629 = 0.064$. In addition, $P(z = 1) = 5556/6812 = 0.816$ and $P[y \in \{1, 2\} | z = 1] = (4332 + 297)/5556 = 0.833$. Hence $r(x)$ defined in equation (3.4b) has the value 0.787; here x is the event $\{y \in \{1, 2\}\}$. Proposition 3.1 now yields this identification region for the official unemployment rate:

$$P[y = 1 | y \in \{1, 2\}] \in [(0.064)(0.787), \ (0.064)(0.787) + 0.213]$$

$$= [0.050, 0.263].$$

Similar computations yield $P[y = 1 | \text{BASE}, y \in \{1, 2\}] \in [0.057, \ 0.164]$.

Complement 3B. Parametric Prediction with Missing Data

Whereas this chapter has studied nonparametric prediction of outcomes conditional on covariates, researchers often specify a parametric family of predictor functions and seek to infer a member of this family that minimizes expected loss with respect to some loss function. Let the outcome y be real-valued. Let Θ be the parameter space and $f(\cdot, \cdot): X \times \Theta \to R$ be the family of predictor functions. Let $L(\cdot): R \to [0, \infty]$ be the loss function. The immediate objective is to find a $\theta^* \in \Theta$ such that

$$\theta^* \in \text{argmin}_{\theta \in \Theta} \ E\{L[y - f(x, \theta)]\}.$$

Then $f(\cdot, \theta^*)$ is called a best $f(\cdot, \cdot)$-predictor of y given x under loss function L. Under usual regularity conditions, θ^* is unique.

For example, consider the familiar problem of best linear prediction under square loss. Here $f(x, \theta) = x'\theta$ and $L[y - f(x, \theta)] = (y - x'\theta)^2$. As is well known, $\theta^* = E(xx')^{-1}E(xy)$, provided that $E(xx')$ is non-singular.

Prediction Ignoring Missing Data

Empirical researchers routinely discard all realizations of (y, x) that are incompletely observed. Suppose that a random sample of size N has been drawn from population J. Let $N(1)$ denote the sample members for which $z_{yx} = 1$, and let N_1 be the cardinality of $N(1)$. Then it is routine to estimate θ^* by a $\theta_N \in \Theta$ such that

$$\theta_N \in \underset{\theta \in \Theta}{\text{argmin}} \quad \frac{1}{N_1} \sum_{i \in N(1)} L[y_i - f(x_i, \theta)].$$

Under usual regularity conditions, θ_N almost surely is unique and

$$\lim_{N \to \infty} \theta_N = \underset{\theta \in \Theta}{\text{argmin}} \; E\{L[y - f(x, \theta)] | z_{yx} = 1\}, \quad \text{a. s.}$$

Thus θ_N is a consistent estimate of θ^* if and only if

$$\underset{\theta \in \Theta}{\text{argmin}} \; E\{L[y - f(x, \theta)] | z_{yx} = 1\} = \underset{\theta \in \Theta}{\text{argmin}} \; E\{L[y - f(x, \theta)]\}.$$

Prediction Using the Empirical Evidence Alone

Consider the problem of parametric prediction using the empirical evidence alone. The identification region for θ^* is the set of parameter values that minimize expected loss under some feasible distribution for the missing data. It is easy enough to characterize this region, but it may be rather difficult to estimate it.

Emulating the construction of $H[P(y | x = x)]$ for general missing-data patterns in Proposition 3.8, the identification region for θ^* is

$$H(\theta^*) = \bigcup_{(\eta_{10}, \eta_{00}, \eta_{01}) \in \Gamma_{10} \times \Gamma_{00} \times \Gamma_{01}}$$

$$\{\underset{\theta \in \Theta}{\text{argmin}} \; P(z_{yx} = 1) \, E\{L[y - f(x, \theta)] | z_{yx} = 1\}$$

$$+ P(z_x = 1, z_y = 0) \cdot \int L[y - f(x, \theta)] d\eta_{10}$$

$$+ P(z_x = 0, z_y = 0) \cdot \int L[y - f(x, \theta)] d\eta_{00}$$

$$+ P(z_x = 0, z_y = 1) \cdot \int L[y - f(x, \theta)] d\eta_{01}\}.$$

Here Γ_{10} is the set of all distributions on $Y \times X$ with x-marginal $P(x | z_x = 1, z_y = 0)$, Γ_{00} is the set of all distributions on $Y \times X$, and Γ_{01} is the set of all distributions on $Y \times X$ with y-marginal $P(y | z_x = 0, z_y = 1)$.

The natural estimate of $H(\theta^*)$ is its sample analog, which uses the empirical distribution of the data to estimate $P(z_{yx})$, $P[(y, x) | z_{yx} = 1]$, $P(x | z_x = 1, z_y = 0)$, and $P(y | z_x = 0, z_y = 1)$. However, computation of this estimate can pose a considerable challenge. This is so even in the relatively benign setting of best linear prediction under square loss, where the sample analog of $H(\theta^*)$ is the set of least squares estimates produced by conjecturing all possible values for missing outcome and covariate data; see Horowitz and Manski (2001).

Prediction Assuming that $f(\cdot, \ \theta^*)$ is Best Nonparametric

Researchers posing parametric prediction problems often combine the empirical evidence with distributional assumptions. In particular, it is common to assume that the best $f(\cdot, \ \cdot)$-predictor of y given x under loss function L is best nonparametric; that is,

$$f(x, \ \theta^*) \ \in \ \text{argmin}_{c \in R} \ E[L(y - c)|x = x], \ \ \forall x \in X.$$

This assumption may enable a researcher to shrink $H(\theta^*)$.

For example, researchers seeking the best linear predictor under square loss often assume that the best nonparametric predictor, the conditional expectation $E(y|x)$, is a linear function of x. Let $H[E(y|x)]$ be the identification region for $E(y|x)$ using the empirical evidence alone. Then the assumption of *linear mean regression* implies that θ is a feasible solution to the problem of best linear prediction under square loss if and only if $x\theta \in H[E(y|x)]$.

Endnotes

Sources and Historical Notes

Much of the analysis in this chapter was originally developed in Horowitz and Manski (1998, 2000). In particular, Propositions 3.1 and 3.5 are based on Horowitz and Manski (1998). Corollary 3.8.1 is based on Horowitz and Manski (2000, Theorem 1).

Whereas prediction with missing outcome data has long been a prominent concern of statistics, serious attention to missing covariate data is relatively recent and far less common. When statisticians have studied missing covariate data, they have invariably imposed assumptions that point-identify $P(y|x)$. Having done this, their main concern has been to understand the finite-sample properties of point estimates of conditional predictors (see, for example, Little, 1992; Robins, Rotnitzky, and Zhao, 1994; Wang, Wang, Zhao, and Ou, 1997).

Text Notes

1. Likewise, interval measurement of real-valued covariates is qualitatively different from interval measurement of outcomes. It was shown in Section 1.5 that, using the empirical evidence alone, interval measurement of outcomes yields simple bounds on parameters that respect stochastic dominance. Corresponding findings when covariates are interval-measured are not available. However, Manski and Tamer (2002) show that imposition of certain distributional assumptions does produce findings of interest.

Let $X \subset R$. Let each $j \in J$ have a triple $(x_{j-}, x_j, x_{j+}) \in X^3$. Let the random variable $(x_-, x, x_+): J \to X^3$ have a distribution $P(x_-, x, x_+)$ such that $P(x_- \leq x \leq x_+) = 1$. Then we have interval measurement of covariates if realizations of (x_-, x_+) are observable but realizations of x are not directly observable. Now suppose that these distributional assumptions hold:

Monotonicity: $E(y|x)$ is weakly increasing in x.

Mean Independence: $E(y|x, x_-, x_+) = E(y|x)$.

Manski and Tamer (2002, Proposition 1) show that, for any $x \in X$,

$$\sup_{(x_-, x_+) \text{ s.t. } x_- \leq x_+ \leq x} [E(y|x_- = x_-, x_+ = x_+)] \leq E(y|x = x) \leq \inf_{(x_-, x_+) \text{ s.t. } x \leq x_- \leq x_+} [E(y|x_- = x_-, x_+ = x_+)].$$

2. Further analysis of conditional prediction with general missing data patterns is presented in Zaffalon (2002). He supposes that the set $Y \times X$ is finite and shows that the smallest and largest feasible values of $E(y|x = x)$ can be obtained by solving fractional linear programming problems.

3. The only finding to date for general missing-data processes concerns functionals of the form $P(y \in B \,|\, x = x) - P(y \in B \,|\, x = x')$, where $B \subset Y$ and where (x, x') are distinct covariate values. Through a laborious derivation, Horowitz and Manski (2000) obtain closed-form expressions for the minimum and maximum feasible values of this functional.

4

Contaminated Outcomes

4.1. The Mixture Model of Data Errors

Throughout the analysis of missing-data problems in Chapters 1 through 3, it was assumed that the available data are realizations of the outcomes and covariates of interest. Researchers use the broad term *data errors* to describe situations in which the available data imperfectly measure variables of interest. In general, data errors produce identification problems. The specific nature of the problem depends on how the available data may be related to the variables of interest.

One prominent conceptualization of data errors has been the *mixture model*, which views the available data as realizations of a probability mixture of the variable of interest and of another random variable. Let each member j of population J have a pair of outcomes (y_j^*, e_j) in the space $Y \times Y$. Let the random variable (y^*, e): $J \to Y \times Y$ have distribution $P(y^*, e)$. Let y^* be the outcome of interest. Let a sampling process draw persons at random from J. The mixture model views the available data as realizations of the probability mixture

$$y \equiv y^*z + e(1 - z), \tag{4.1}$$

where z is an unobservable binary random variable indicating whether e or y^* is observed; Thus y^* is observed if $z = 1$ and e is observed if $z = 0$. Realizations of y with $z = 0$ are said to be data errors, those with $z = 1$ are said to be *error-free*, and y itself is said to be a *contaminated* measure of y^*.

The mixture model has no content per se but may be informative about y^* when combined with distributional assumptions. Researchers often

60

assume that the *error probability* $P(z = 0)$ is known, or at least that it can be bounded non-trivially from above. This chapter studies identification of outcome distributions under such assumptions.

Let $p \equiv P(z = 0)$ denote the probability of a data error. The inferential problem is displayed by the Law of Total Probability in equations (4.2) and (4.3):

$$P(y) = P(y|z = 1)(1 - p) + P(y|z = 0)p \qquad (4.2)$$

and

$$P(y^*) = P(y|z = 1)(1 - p) + P(y^*|z = 0)p. \qquad (4.3)$$

The sampling process reveals only the distribution $P(y)$ on the left side of (4.2). Empirical knowledge of $P(y)$ per se is uninformative about $P(y|z = 1)$ and, hence, about $P(y^*)$. However, informative identification regions emerge if knowledge of $P(y)$ is combined with a non-trivial upper bound on p.

An upper bound on the probability of data errors has been a central assumption of research on robust inference in the presence of data errors (see Complement 4B). In some applications, the probability of a data error can be estimated from a validation data set. In other applications, data errors arise out of the efforts of analysts to impute missing values; the fraction of imputed values then provides the upper bound on the error probability (see Complement 4A). There are, of course, many applications in which there is no obvious way to set a firm upper bound on the probability of a data error. In these cases, it may still be of interest to determine how inference on population parameters degrades as the error probability increases.

This chapter studies the identification of two outcome distributions. One is the distribution

$$P(y|z = 1) \equiv P(y^*|z = 1) \qquad (4.4)$$

of error-free realizations. The other is the marginal distribution $P(y^*)$ of the outcome of interest. These two distributions generally are distinct, but they are identical if the occurrence of data errors is statistically independent of the outcome of interest. Thus, the chapter effectively presents two parallel sets of findings on identification of $P(y^*)$. One set of findings assumes only an upper bound on the error probability, and the other also assumes that

$$P(y^*) = P(y^*|z = 1). \qquad (4.5)$$

Section 4.2 develops the identification regions for $P(y|z=1)$ and $P(y^*)$. These abstract findings are fleshed out in Sections 4.3 and 4.4, which derive simple identification regions for event probabilities and for parameters that respect stochastic dominance. All results apply to inference on conditional distributions $P(y|x=x, z=1)$ and $P(y^*|x=x)$ if the covariates x are known to be measured without error; one simply redefines the population of interest to be the sub-population for which $\{x=x\}$. The present analysis does not cover cases in which covariates are measured with error.

Considered abstractly, this chapter studies identification of the components of a probability mixture. Contaminated outcomes is only one of many manifestations of this basic identification problem. In Chapter 3, the problem appeared in the analysis of missing covariate data. In Chapters 5 and 10, it will arise when we study *ecological inference* and the *mixing problem* of program evaluation.

4.2. Outcome Distributions

Part (a) of Proposition 4.1 shows that, if p is known, $P(y|z=1)$ belongs to the identification region $H_p[P(y|z=1)]$ defined below and $P(y^*)$ belongs to a larger region $H_p[P(y^*)]$. Part (b) shows that these regions expand as the error probability increases. This implies, in part (c), that the identification regions given an upper bound, say λ, on p are $H_\lambda[P(y|z=1)]$ and $H_\lambda[P(y^*)]$.

Proposition 4.1: (a) Let p be known, with $p < 1$. Then the identification regions for $P(y|z=1)$ and $P(y^*)$ are

$$H_p[P(y|z=1)] \equiv \Gamma_Y \cap \{[P(y) - p\gamma]/(1-p), \gamma \in \Gamma_Y\} \qquad (4.6)$$

and

$$H_p[P(y^*)] \equiv \{(1-p)\eta + p\gamma, (\eta, \gamma) \in H_p[P(y|z=1)] \times \Gamma_Y\}. \qquad (4.7)$$

(b) Let $\delta > 0$ and $p + \delta < 1$. Then $H_p[P(y|z=1)] \subset H_{p+\delta}[P(y|z=1)]$ and $H_p[P(y^*)] \subset H_{p+\delta}[P(y^*)]$.

(c) For given $\lambda < 1$, let it be known that $p \leq \lambda$. Then the identification regions for $P(y|z=1)$ and $P(y^*)$ are $H_\lambda[P(y|z=1)]$ and $H_\lambda[P(y^*)]$. $\square$

Proof: (a) By (4.2), the joint identification region for $[P(y|z=1), P(y|z=0)]$ is

$$H_p[P(y|z=1), P(y|z=0)] \equiv \{(\eta, \gamma) \in \Gamma_Y \times \Gamma_Y: P(y) = (1-p)\eta + p\gamma\}.$$

Equation (4.6) follows immediately. Equation (4.7) follows from (4.3) because the sampling process is uninformative about $P(y^*|z=0)$.

(b) To show that $H_p[P(y|z=1)] \subset H_{p+\delta}[P(y|z=1)]$, consider $(\eta, \gamma) \in H_p[P(y|z=1), P(y|z=0)]$. Let the error probability increase from p to p+δ. Let $\gamma_\delta \equiv (\gamma p + \eta\delta)/(p+\delta)$. Then γ_δ is a probability distribution and $(\eta, \gamma_\delta) \in \Gamma_Y \times \Gamma_Y$ solves the equation

$$P(y) = \eta(1-p-\delta) + \gamma_\delta(p+\delta).$$

Hence $(\eta, \gamma_\delta) \in H_{p+\delta}[P(y|z=1), P(y|z=0)]$.
   That $H_p[P(y^*)] \subset H_{p+\delta}[P(y^*)]$ follows from the result above and (4.7).

(c) The identification region for $P(y|z=1)$ is $\cup_{p\le\lambda} H_p[P(y|z=1)]$. Part (b) showed that $\cup_{p\le\lambda} H_p[P(y|z=1)] = H_\lambda[P(y|z=1)]$. Similarly, the identification region for $P(y^*)$ is $\cup_{p\le\lambda}H_p[P(y^*)]$, and part (b) showed that $\cup_{p\le\lambda}H_p[P(y^*)] = H_\lambda[P(y^*)]$.

Q. E. D.

Observe that, whatever the error probability p may be, the distribution $P(y)$ of observed outcomes necessarily belongs to $H_p[P(y|z=1)]$. Hence the hypothesis $\{P(y^*) = P(y|z=1) = P(y)\}$ is not refutable.

4.3. Event Probabilities

Let B be a measurable subset of Y. Proposition 4.1 implies simple identification regions for the event probabilities $P(y \in B|z=1)$ and $P(y^* \in B)$. Proposition 4.2 derives these regions. The proposition shows that there are informative lower bounds on both $P(y \in B|z=1)$ and $P(y^* \in B)$ if $P(y \in B) > \lambda$ and informative upper bounds if $P(y \in B) \le 1 - \lambda$. Thus $\lambda < \frac{1}{2}$ is a necessary condition for there to be both informative lower and upper bounds on event probabilities.

Proposition 4.2: (a) Let p be known, with p < 1. Then the identification regions for $P(y \in B|z=1)$ and $P(y^* \in B)$ are the intervals

$$H_p[P(y \in B|z=1)] = [0, 1] \cap [[P(y \in B) - p]/(1-p), P(y \in B)/(1-p)],$$

(4.8)

$$H_p[P(y^* \in B)] = [0, 1] \cap [P(y \in B) - p, P(y \in B) + p]. \qquad (4.9)$$

(b) For given $\lambda < 1$, let it be known that $p \le \lambda$. Then the identification regions for $P(y \in B | z = 1)$ and $P(y^* \in B)$ are $H_\lambda[P(y | z = 1)]$ and $H_\lambda[P(y^*)]$.   □

Proof: (a) The first task is to prove that (4.8) gives the identification region for $P(y \in B | z = 1)$. Proposition 4.1 implies that $P(y \in B | z = 1)$ lies in the interval

$$[0, 1] \cap \{[P(y \in B) - pa]/(1 - p), a \in [0, 1]\} =$$

$$[0, 1] \cap [[P(y \in B) - p]/(1 - p), P(y \in B)/(1 - p)].$$

Thus, the identification region for $[P(y \in B | z = 1)]$ is a subset of the set on the right side of (4.8). We must prove that all elements of this set are feasible. This is so if, for all $c \in H_p[P(y \in B | z = 1)]$, there exist distributions $(\eta, \gamma) \in \Gamma_Y \times \Gamma_Y$ such that $\eta(B) = c$ and $P(y) = (1 - p)\eta + p\gamma$.

To prove that such distributions exist, let $c \in H_p[P(y \in B | z = 1)]$ and let d solve the equation

$$P(y \in B) = (1 - p)c + pd.$$

Equation (4.8) implies that $d \in [0, 1]$. Now choose (η, γ) as follows:
If $P(y \in B) > 0$ and $A \subset B$, let
$\qquad \eta(A) = [P(y \in A)/P(y \in B)]c,$
$\qquad \gamma(A) = [P(y \in A)/P(y \in B)]d.$
If $P(y \in B) = 0$ and $A \subset B$, let $\eta(A) = \gamma(A) = 0$.
If $P(y \in Y - B) > 0$ and $A \subset Y - B$, let
$\qquad \eta(A) = [P(y \in A)/P(y \in Y - B)](1 - c),$
$\qquad \gamma(A) = [P(y \in A)/P(y \in Y - B)](1 - d).$
If $P(y \in Y - B) = 0$ and $A \subset Y - B$, let $\eta(A) = \gamma(A) = 0$.
Then $\eta(B) = c$ and $P(y \in A) = (1 - p)\eta(A) + p\gamma(A)$ for all $A \subset B$ and for all $A \subset Y - B$. Hence $P(y) = (1 - p)\eta + p\gamma$.

The second task is to prove that (4.9) gives the identification region for $P(y^* \in B)$. The sampling process is uninformative about $P(y^* \in B | z = 0)$. Hence the identification region for $P(y^* \in B)$ is the set

$$\{(1 - p)c + pa, \quad c \in H_p[P(y \in B | z = 1)], a \in [0, 1]\}$$

$$= \{[0, 1 - p] \cap [P(y \in B) - p, P(y \in B)] + p[0, 1]\}$$

$$= [0, 1] \cap [P(y \in B) - p, P(y \in B) + p].$$

(b) The identification regions for $P(y \in B | z = 1)$ and $P(y^* \in B)$ are $\cup_{p \leq \lambda} H_p[P(y \in B | z = 1)]$ and $\cup_{p \leq \lambda} H_p[P(y^* \in B)]$. It follows from part (a) that

$$\cup_{p \leq \lambda} H_p[P(y \in B | z = 1)] = H_\lambda[P(y \in B | z = 1)]$$

$$\cup_{p \leq \lambda} H_p[P(y^* \in B)] = H_\lambda[P(y^* \in B)].$$

Q. E. D.

4.4. Parameters that Respect Stochastic Dominance

Suppose now that the outcome y is real-valued. Proposition 4.3 shows that, for each $p \in [0, 1]$, the identification region $H_p[P(y|z = 1)]$ contains a "smallest" member L_p that is stochastically dominated by all feasible values of $P(y|z = 1)$ and a "largest" member U_p that stochastically dominates all feasible values of $P(y|z = 1)$. These smallest and largest distributions are truncated versions of the distribution $P(y)$ of observed outcomes: L_p right-truncates $P(y)$ at its $(1-p)$–quantile and U_p left-truncates $P(y)$ at its p–quantile. Proposition 4.3 uses distributions L_p and U_p to determine the smallest and largest feasible values for parameters that respect stochastic dominance.

Proposition 4.3: Let Y be a subset of R that contains its lower and upper bounds, y_0 and y_1. Let $D(\cdot)$ respect stochastic dominance. For $\alpha \in [0, 1]$, let $Q_\alpha(y)$ denote the α–quantile of $P(y)$. Define probability distributions L_α and U_α on R as follows:

$$L_\alpha[-\infty, t] \equiv P(y \leq t)/(1 - \alpha) \qquad \text{for } t < Q_{(1-\alpha)}(y)$$
$$\equiv 1 \qquad \text{for } t \geq Q_{(1-\alpha)}(y). \qquad (4.10a)$$

$$U_\alpha[-\infty, t] \equiv 0 \qquad \text{for } t < Q_\alpha(y)$$
$$\equiv [P(y \leq t) - \alpha]/(1 - \alpha) \qquad \text{for } t \geq Q_\alpha(y). \qquad (4.10b)$$

Let $\gamma_0 \in \Gamma_Y$ and $\gamma_1 \in \Gamma_Y$ be the degenerate distributions placing all mass on y_0 and y_1 respectively.

(a) Let p be known, with $p < 1$. Then sharp lower and upper bounds on $D[P(y|z = 1)]$ are $D(L_p)$ and $D(U_p)$. Sharp bounds on $D[P(y^*)]$ are $D[(1 - p)L_p + p\gamma_0]$ and $D[(1 - p)U_p + p\gamma_1]$.

(b) For given $\lambda < 1$, let it be known that $p \leq \lambda$. Then sharp lower and upper bounds on $D[P(y|z = 1)]$ are $D(L_\lambda)$ and $D(U_\lambda)$. Sharp bounds on $D[P(y^*)]$ are

$D[(1 - \lambda)L_\lambda + \lambda\gamma_0]$ and $D[(1 - \lambda)U_\lambda + \lambda\gamma_1]$.   □

Proof: (a) I first show that $D(L_p)$ is the sharp lower bound on $D[P(y|z=1)]$. $D(L_p)$ is a feasible value for $D[P(y|z=1)]$ because

$$P(y \le t) = (1 - p)L_p[-\infty, t] + pU_{(1-p)}[-\infty, t], \quad \forall\, t \in R.$$

Thus $(L_p, U_{(1-p)}) \in H_p[P(y|z=1), P(y|z=0)]$. $D(L_p)$ is the smallest feasible value for $D[P(y|z=1)]$ because L_p is stochastically dominated by every member of $H_p[P(y|z=1)]$. To prove this, one needs to show that $L_p[-\infty, t]$ $\ge \eta[-\infty, t]$ for all $\eta \in H_p[P(y|z=1)]$ and all $t \in R$. Fix η. If $t \ge Q_{(1-p)}(y)$, then

$$L_p[-\infty, t] - \eta[-\infty, t] = 1 - \eta[-\infty, t] \ge 0.$$

If $t < Q_{(1-p)}(y)$, then

$$\eta[-\infty, t] > L_p[-\infty, t] \Rightarrow (1 - p)\eta[-\infty, t] > P(y \le t)$$

$$\Rightarrow (1 - p)\eta[-\infty, t] + p\gamma[-\infty, t] > P(y \le t)$$

for all $\gamma \in \Gamma_Y$. This contradicts the supposition that $\eta \in H_p[P(y|z=1)]$, so $\eta[-\infty, t] \le L_p[-\infty, t]$ for all t.

Now consider $D(U_p)$. This is a feasible value for $D[P(y|z=1)]$ because

$$P(y \le t) = (1 - p)U_p[-\infty, t] + pL_{(1-p)}[-\infty, t], \quad \forall\, t \in R.$$

Thus $(U_p, L_{(1-p)}) \in H_p[P(y|z=1), P(y|z=0)]$. $D(U_p)$ is the largest feasible value for $D[P(y|z=1)]$ because U_p stochastically dominates every member of $H_p[P(y|z=1)]$. To prove this, one needs to show that $U_p[-\infty, t] \le \eta[-\infty, t]$ for all $\eta \in H_p[P(y|z=1)]$ and all $t \in R$. Fix η. If $t < Q_p(y)$, then

$$U_p[-\infty, t] - \eta[-\infty, t] = 0 - \eta[-\infty, t] \le 0.$$

If $t \ge Q_p(y)$, then

$$\eta[-\infty, t] < U_p[-\infty, t] \Rightarrow (1 - p)\eta[-\infty, t] < P(y \le t) - p$$

$$\Rightarrow (1 - p)\eta[-\infty, t] + p\gamma[-\infty, t] < P(y \le t)$$

for all $\gamma \in \Gamma_Y$. This contradicts the supposition that $\eta \in H_p[P(y|z=1)]$, so $U_p[-\infty, t] \le \eta[-\infty, t]$ for all t.

Now consider $D[P(y^*)]$. By Proposition 4.1, the identification region for $P(y^*)$ is

$$H_p[P(y^*)] \equiv \{(1 - p)\eta + p\gamma, (\eta, \gamma) \in H_p[P(y|z = 1)] \times \Gamma_Y\}.$$

We have found that $L_p \in H_p[P(y|z = 1)]$ and that L_p is stochastically dominated by all members of $H_p[P(y|z = 1)]$. Distribution γ_0 belongs to Γ_Y and is stochastically dominated by all members of Γ_Y. Hence distribution $(1 - p)L_p + p\gamma_0$ belongs to $H_p[P(y^*)]$ and is stochastically dominated by all members of $H_p[P(y^*)]$. Hence $D[(1 - p)L_p + p\gamma_0]$ is the smallest feasible value for $D[P(y^*)]$. The proof for the upper bound is analogous.

(b) By part (a), the sharp lower and upper bounds on $D[P(y|z = 1)]$ are $\inf_{p \le \lambda} D(L_p)$ and $\sup_{p \le \lambda} D(U_p)$. Consider the lower bound. Part (b) of Proposition 4.1 and part (a) of the present proposition imply that $L_p \in H_\lambda[P(y|z = 1)], \forall p \le \lambda$. Moreover, part (a) showed that L_λ is stochastically dominated by all elements of $H_\lambda[P(y|z = 1)]$. Hence $\inf_{p \le \lambda} D(L_p) = D(L_\lambda)$. The same reasoning shows that $\sup_{p \le \lambda} D(U_p) = D(U_\lambda)$. Given these results, the proof for $D[P(y^*)]$ is the same as in part (a).

Q. E. D.

Quantiles
Proposition 4.3 yields simple sharp lower and upper bounds on quantiles of $P(y|z = 1)$ and $P(y^*)$. Corollary 4.3.1 shows that the bounds on quantiles of $P(y|z = 1)$ are informative whenever the error probability is known to be less than one. However, the bounds on quantiles of $P(y^*)$ are informative only if the error probability is sufficiently small. For $\alpha \in (0, 1)$, there is an informative lower bound on the α–quantile of $P(y^*)$ only if $\alpha > \lambda$ and an informative upper bound only if $\alpha \le 1 - \lambda$.

Corollary 4.3.1: Let Y be a subset of R that contains its lower and upper bounds, y_0 and y_1. Let $\alpha \in (0, 1)$. For $a \in R$, define

$$r_a(y) \equiv Q_a(y) \text{ if } 0 < a < 1; \quad r_a(y) \equiv y_0 \text{ if } a \le 0; \quad r_a(y) \equiv y_1 \text{ if } a \ge 1.$$

(a) Let p be known, with $p < 1$. Then sharp lower and upper bounds on the α–quantile of $P(y|z = 1)$ are $Q_{\alpha(1-p)}(y)$ and $Q_{[\alpha(1-p)+p]}(y)$. Sharp bounds on the α–quantile of $P(y^*)$ are $r_{(\alpha-p)}(y)$ and $r_{(\alpha+p)}(y)$.

(b) For given $\lambda < 1$, let it be known that $p \le \lambda$. Then sharp lower and upper bounds on the α–quantile of $P(y|z = 1)$ are $Q_{\alpha(1-\lambda)}(y)$ and $Q_{[\alpha(1-\lambda)+\lambda]}(y)$. Sharp bounds on the α–quantile of $P(y^*)$ are $r_{(\alpha-\lambda)}(y)$ and $r_{(\alpha+\lambda)}(y)$. □

Proof: (a) Proposition 4.3 showed that the smallest and largest feasible values for the α–quantile of $P(y|z = 1)$ are the α–quantiles of L_p and U_p, which are $Q_{\alpha(1-p)}(y)$ and $Q_{[\alpha(1-p)+p]}(y)$. Proposition 4.3 showed that the smallest and largest feasible values for the α–quantile of $P(y^*)$ are the α–quantiles of $[(1 - p)L_p + p\gamma_0]$ and $[(1 - p)U_p + p\gamma_1]$, which are $r_{(\alpha-p)}(y)$ and $r_{(\alpha+p)}(y)$.

(b) This is an immediate application of Proposition 4.3, part (b).

$\hfill$ Q. E. D.

Means

Proposition 4.3 yields simple sharp lower and upper bounds on the expectations $E(y|z = 1)$ and $E(y^*)$. Corollary 4.3.2 shows that the bounds on $E(y|z = 1)$ are informative whenever the error probability is known to be less than one. The lower bound on $E(y^*)$ is informative if y_0 is finite, and the upper bound is informative if y_1 is finite. These results are immediate.

Corollary 4.3.2: Let Y be a subset of R that contains its lower and upper bounds, y_0 and y_1.

(a) Let p be known, with $p < 1$. Then sharp lower and upper bounds on $E(y|z = 1)$ are $\int y dL_p$ and $\int y dU_p$. Sharp bounds on $E(y^*)$ are $(1-p)\int y dL_p + py_0$ and $(1-p)\int y dU_p + py_1$.

(b) For given $\lambda < 1$, let it be known that $p \le \lambda$. Then sharp lower and upper bounds on $E(y|z = 1)$ are $\int y dL_\lambda$ and $\int y dU_\lambda$. Sharp bounds on $E(y^*)$ are $(1 - \lambda)\int y dL_\lambda + \lambda y_0$ and $(1 - \lambda)\int y dU_\lambda + \lambda y_1$. $\hfill \square$

Complement 4A. Contamination Through Imputation

Organizations conducting major surveys often impute values for missing data and report statistics that mix real and imputed data. This practice may, but need not, yield consistent estimates of population quantities of interest.

Consider an observer who sees the reported statistics but who does not see the raw survey data and does not know the imputation rule used when data are missing. To this observer, imputations are data errors that may be analyzed using the findings of this chapter. I use income statistics published by the U.S. Bureau of the Census to illustrate.

Income Distribution in the United States

Data on the household income distribution in the United States are collected annually in the Current Population Survey (CPS). Summary statistics are published by the U.S. Bureau of the Census (the Bureau) in Series P-60 of its Current Population Reports. Two sampling problems identified by the Bureau are *interview nonresponse*, wherein some households in the CPS sampling frame are not interviewed, and *item nonresponse*, wherein some of those interviewed do not provide complete income responses. Faced with these nonresponse problems, the Bureau uses available information to impute missing income data. The Bureau mixes actual and imputed data to produce the household income statistics reported in its Series P-60 publications.

From the perspective of this chapter, y^* is the income a household selected for interview in the CPS would report if it were to complete the survey, e is the income the Bureau would impute to the household if the household were not to complete the survey, and $z = 1$ if a CPS household actually completes the survey. $P(y|z = 1)$ is the distribution of income reported by those CPS households who complete the survey, $P(y^*)$ is the distribution of income that would be reported if all households in the CPS sampling frame were to complete the survey, and $P(y)$ is the distribution of household income found in the Series P-60 publications. The error probability p is the probability that a CPS household does not complete the survey.

The Bureau's imputation practice is valid if the distribution $P(y|z = 0)$ of incomes imputed for persons who do not complete the survey coincides with the distribution $P(y^*|z = 0)$ that these persons would report if they were to complete the survey; then $P(y) = P(y^*)$. However, $P(y|z = 0)$ and $P(y^*|z = 0)$ could differ markedly. The identification regions developed in this chapter are agnostic about the quality of the Bureau imputation practice.

Consider the year 1989. U.S. Bureau of the Census (1991, pp. 387–388) states that in the March 1990 CPS, which provides data on incomes during 1989, approximately 4.5% of the 60,000 households in the sampling frame were not interviewed and that incomplete income data were obtained from approximately 8% of the persons in interviewed households. The Bureau's publication does not report how the latter group are spread across households but we can be sure that no more than $(0.08)(0.955) = 0.076$ of the households have item nonresponse, so $\lambda = 0.121$ provides an upper bound on p.

Now consider $P(y)$. U.S. Bureau of the Census (1991, Table 5, p. 17) provides findings for each of twenty-one income intervals (in thousands of dollars):

P[0, 5) = 0.053	P[35, 40) = 0.066	P[70, 75) = 0.018
P[5, 10) = 0.103	P[40, 45) = 0.060	P[75, 80) = 0.015
P[10, 15) = 0.097	P[45, 50) = 0.048	P[80, 85) = 0.013
P[15, 20) = 0.092	P[50, 55) = 0.043	P[85, 90) = 0.009
P[20, 25) = 0.087	P[55, 60) = 0.032	P[90, 95) = 0.008
P[25, 30) = 0.083	P[60, 65) = 0.028	P[95, 100) = 0.006
P[30, 35) = 0.076	P[65, 70) = 0.023	P[100, ∞) = 0.039

Let us "fill out" P(y) by imposing the auxiliary assumption that income is distributed uniformly within each interval except the last. We may now obtain bounds on features of $P(y|z = 1)$ and $P(y^*)$.

For example, consider the probability that household income is below $30,000. We have P[0,30) = 0.515 and λ = 0.121. Hence, the bound on $P(y \leq 30|z = 1)$ is [0.448, 0.586] and the bound on $P(y^* \leq 30)$ is [0.394, 0.636]. Now consider median household income. The median of $P(y|z = 1)$ must lie between the $0.5(1 - \lambda)$ and $[0.5(1 - \lambda) + \lambda]$–quantiles of P(y), while the median of $P(y^*)$ must lie between the $(0.5 - \lambda)$ and $(0.5 + \lambda)$–quantiles of P(y). Invoking the auxiliary assumption that P(y) is uniform within $5000 intervals, the sharp lower and upper bounds on the median of $P(y|z = 1)$ are [25.482, 33.026], and the corresponding bounds on the median of $P(y_1)$ are [21.954, 37.273].

These bounds illustrate what one can learn about the distribution of income using the statistics reported in the Series P-60 publications. Tighter inferences can be drawn if one has access to the raw CPS survey data, which flag the cases in which income data are imputed. Access to the raw data enables one to point-identify $P(y|z = 1)$ and also to learn the error probability p. With this information, one faces a problem of missing outcomes rather than the more severe problem of contaminated outcomes.

Complement 4B. Identification and Robust Inference

The mixture model of data errors has long been a central concern of research on *robust inference*. Huber (1964) combined the mixture model with knowledge of an upper bound on the error probability to develop minimax estimators of location parameters. The literature on robust inference that has developed out of Huber's work has not sought to determine identification regions for population parameters. Rather, it has aimed to characterize how point estimates of population parameters behave when data errors are generated in specified ways. The main objective has been to find point estimates that are not greatly affected by errors. Huber (1981) and Hampel, Ronchetti, Rousseeuw, and Stahel (1986) present comprehensive treatments of robust inference.

The pre-occupation of robust inference with point estimation stands in contrast to the perspective of this book. In general, I find it difficult to motivate point estimation of parameters that are only partially identified. It seems to me more natural to estimate the identification regions for such parameters, or at least their sharp lower and upper bounds.

Although the literature on robust inference has focused on point estimation, in other respects it has been more conservative than identification analysis. The practice in robustness studies has been to consider the inference problem before data are collected. The objective is to guard against the worst outcomes that errors in the data could conceivably produce. But some outcomes that are possible *ex ante* can be ruled out *ex post*—after the data have been collected. Identification analysis characterizes the inferences that can be made given knowledge of the empirical distribution of the available data.

The problem of inferring the mean $E(y|z = 1)$ of the error-free data provides a compelling example of the difference between identification analysis and robust inference. It is well-known that $E(y|z = 1)$ is not robust under the mixture model of data errors. Yet Corollary 4.3.2 of this chapter has determined informative sharp bounds on $E(y|z = 1)$.

These findings are not contradictory. Identification analysis and robust inference take different positions on the available empirical evidence. Identification analysis shows what values of $E(y|z = 1)$ are feasible given the empirical knowledge of $P(y)$ revealed by the sampling process. Thus, Corollary 4.3.2 showed that, given an upper bound λ on the error probability, the range of feasible values of $E(y|z = 1)$ is $[\int y dL_\lambda, \int y dU_\lambda]$. In contrast, robust inference considers the *ex ante* situation in which $P(y)$ is not yet known because the sampling process has not yet been executed. In this setting, robustness studies conjecture some value for $E(y|z = 1)$ and ask what values of $E(y)$ are consistent with this conjecture. Given λ and a conjecture for $E(y|z = 1)$, the set of feasible values for $E(y)$ is

$$\{(1 - \lambda)E(y|z = 1) + \lambda a, \quad a \in [y_0, y_1]\}.$$

This set, which has the same structure as the identification region for $E(y)$ when outcome data are missing, has finite range only if y_0 and y_1 are finite.

Endnotes

Sources and Historical Notes

The analysis in this chapter originally appeared in Horowitz and Manski (1995). In particular, Propositions 4.1 through 4.3 here are based on Propositions 1 and 4 there. Subsequently, Horowitz and Manski (1997) summarized the main findings, explained how the bounds on certain parameters can be estimated, obtained asymptotic confidence regions for the bounds, and showed how to test hypotheses about unidentified population parameters.

The mixture model studied in this chapter is one of two prominent conceptualizations of data errors. The other is the *convolution model*, which views the available data as realizations of a convolution of the outcome of interest and of another random variable u; thus $y \equiv y^* + u$. The observable outcome y is said to measure the unobservable y^* with *errors-in-variables*.

Like the mixture model, the convolution model has no content per se but may be informative when combined with distributional assumptions. Researchers using the convolution model commonly maintain the assumption that u is statistically independent of y^* and that $P(u)$ is centered at zero in some specified sense. The problem of identification of $P(y^*)$ given knowledge of $P(y)$ is then called the *deconvolution problem*.

In general, researchers using the mixture model and the convolution model make non-nested assumptions about the nature of data errors. From the perspective of the convolution model, the error probability of the mixture model is $p = P(u \neq 0)$. From the perspective of the mixture model, the additive error of the convolution model is $u \equiv (e - y^*)(1 - z)$. Researchers using the convolution model generally do not place an a priori upper bound on $P(u \neq 0)$. Researchers using the mixture model generally do not assume that the quantity $(e - y^*)(1 - z)$ is statistically independent of y^*.

5

Regressions, Short and Long

5.1. Ecological Inference

The *ecological inference* problem has long engaged social scientists who aim to predict outcomes conditional on covariates. Let each member j of population J have an outcome y_j in a space Y and covariates (x_j, z_j) in a space $X \times Z$. Let the random variable $(y, x, z): J \to Y \times X \times Z$ have distribution $P(y, x, z)$. The general goal is to learn the conditional distributions $P(y|x, z) \equiv \{P(y|x = x, z = z), (x, z) \in X \times Z\}$. When y is real-valued, a particular objective may be to learn the mean regression $E(y|x, z) \equiv \{E(y|x = x, z = z), (x, z) \in X \times Z\}$.

Suppose that joint realizations of (y, x, z) are not observable. Instead, data are available from two sampling processes. One process draws persons at random from J and generates observable realizations of (y, x) but not z. The other sampling process draws persons at random and generates observable realizations of (x, z) but not y. The two sampling processes reveal the distributions $P(y, x)$ and $P(x, z)$. Ecological inference is the use of this empirical evidence to learn about $P(y|x, z)$.

A prominent example arises in the analysis of voting behavior. A researcher may want to predict voting behavior (y) conditional on electoral district (x) and demographic attributes (z). The available data may include administrative records on voting by district and census data describing the demographic attributes of persons in each district. Voting records reveal $P(y, x)$, and census data reveal $P(z, x)$. However, it may be that no data source reveals $P(y, z, x)$.

This chapter studies the identification problem manifest in ecological inference. To simplify the presentation, it is supposed throughout that $X \times Z$

73

is finite and that $P(x = x, z = z) > 0$ for all $(x, z) \in X \times Z$. This regularity condition is maintained without further reference.

5.2. Anatomy of the Problem

The structure of the ecological inference problem is displayed by the Law of Total Probability

$$P(y|x) = \sum_{z \in Z} P(y|x, z = z)P(z|x). \tag{5.1}$$

The available empirical evidence reveals the *short* conditional distributions $P(y|x)$ and $P(z|x)$, where *short* means that these distributions condition on x but not on z. The objective is inference on the *long* conditional distributions $P(y|x, z)$, where *long* means that these distributions condition on (x, z).

Let $x \in X$. Define $P(y|x = x, z) \equiv [P(y|x = x, z = z), z \in Z]$. Let $|Z|$ be the cardinality of Z. A $|Z|$-vector of distributions $[\eta_z, z \in Z] \in (\Gamma_Y)^{|Z|}$ is a feasible value for $P(y|x = x, z)$ if and only if it solves the finite mixture problem

$$P(y|x = x) = \sum_{z \in Z} \eta_z P(z = z|x = x). \tag{5.2}$$

Hence, the identification region for $P(y|x = x, z)$ using the empirical evidence alone is

$$H[P(y|x = x, z)] =$$

$$\{(\eta_z, z \in Z) \in (\Gamma_Y)^{|Z|} : P(y|x = x) = \sum_{z \in Z} \eta_z P(z = z|x = x)\}. \tag{5.3}$$

Moreover, the identification region for $P(y|x, z)$ is the Cartesian product $\times_{x \in X} H[P(y|x = x, z)]$. This holds because the Law of Total Probability (5.1) only restricts $P(y|x, z)$ across values of z, not across values of x.

Inference on One Long Conditional Distribution

The finite mixture problem stated in equation (5.2) generalizes the binary mixture problem studied in Chapter 4. There the random variable z took the values 0 and 1, indicating a data error and an error-free realization. Here the random variable z takes values in the finite covariate space Z.

Proposition 5.1 shows that the generalization from binary to finite mixtures is inconsequential if the objective is inference on $P(y|x = x, z = z)$

for any one specified covariate value $(x, z) \in X \times Z$. The present identification region for $P(y|x = x, z = z)$ has the same form as the region obtained for the distribution of error-free data $P(y|z = 1)$ in part (a) of Proposition 4.1.

Proposition 5.1: Let $(x, z) \in X \times Z$. Let $p \equiv P(z \neq z|x = x)$. Then the identification region for $P(y|x = x, z = z)$ is

$$H[P(y|x = x, z = z)] = \Gamma_Y \cap \{[P(y|x = x) - p\gamma]/(1-p), \gamma \in \Gamma_Y\}. \quad (5.4)$$

$\square$

Proof: By (5.3), $(\eta_z, z \in Z) \in H[P(y|x = x, z)]$ if and only if $(\eta_z, z \in Z) \in (\Gamma_Y)^{|Z|}$ and

$$\eta_z = [P(y|x = x) - \sum_{z' \in Z, z' \neq z} \eta_{z'} P(z = z'|x = x)]/(1 - p)$$

$$= [P(y|x = x) - p\gamma]/(1 - p),$$

where $\gamma \equiv \sum_{z' \in Z, z' \neq z} \eta_{z'} P(z = z'|x = x)/p$. This shows that all elements of the identification region for $P(y|x = x, z = z)$ belong to the set of distributions on the right side of (5.4).

To show that every member of this set of distributions is feasible, let η_z belong to this set. Then there exists a $\gamma \in \Gamma_Y$ such that

$$P(y|x = x) = (1 - p)\eta_z + p\gamma.$$

Let $\eta_{z'} = \gamma$ for all $z' \in Z, z' \neq z$. Then

$$(1 - p)\eta_z + p\gamma = (1 - p)\eta_z + \sum_{z' \in Z, z' \neq z} \eta_{z'} P(z = z'|x = x).$$

Hence $(\eta_z, z \in Z) \in H[P(y|x = x, z)]$.

Q. E. D.

Joint Inference on Long Conditional Distributions

The generalization from binary to finite mixtures is consequential if the objective is joint inference on the vector of long conditional distributions $P(y|x = x, z)$. This vector of distributions must solve the mixture problem stated in equation (5.2). Hence the identification region $H[P(y|x = x, z)]$ stated in equation (5.3) necessarily is a proper subset of the Cartesian product $\times_{z \in Z} H[P(y|x = x, z = z)]$.

Equation (5.3) is simple in form but is too abstract to communicate much about the size and shape of $H[P(y|x = x, z)]$. Section 5.3 addresses an important aspect of this question, this being the structure of the identification region for long mean regressions. Section 5.4 examines the identifying power of two distributional assumptions using instrumental variables.

5.3. Long Mean Regressions

Let $x \in X$. Equation (5.3) implies that the feasible values of $E(y|x = x, z)$ are

$$H[E(y|x = x, z)] = \{(\int y d\eta_z, z \in Z), (\eta_z, z \in Z) \in H[P(y|x = x, z)]\}. \quad (5.5)$$

This section characterizes $H[E(y|x = x, z)]$ less abstractly than (5.5).

Some Immediate Properties

Some properties of $H[E(y|x = x, z)]$ are immediate. First, observe that the set $H[P(y|x = x, z)]$ is convex and the expectation operator is linear. Hence $H[E(y|x = x, z)]$ is a convex set.

Second, observe that, for each $z \in Z$, we already have sharp bounds on $E(y|x = x, z = z)$. For $a \in [0, 1]$, let $Q_a(y|x = x)$ denote the a–quantile of distribution $P(y|x = x)$ and define

$$L_a[-\infty, t] \equiv P(y \leq t|x = x)/(1 - a) \qquad \text{for } t < Q_{(1-a)}(y|x = x)$$
$$\equiv 1 \qquad\qquad\qquad\qquad \text{for } t \geq Q_{(1-a)}(y|x = x) \qquad (5.6a)$$

$$U_a[-\infty, t] \equiv 0 \qquad\qquad\qquad\qquad\qquad \text{for } t < Q_a(y|x = x)$$
$$\equiv [P(y \leq t|x = x) - a]/(1 - a) \quad \text{for } t \geq Q_a(y|x = x). \qquad (5.6b)$$

Let L_{xz} and U_{xz} be the distributions L_a and U_a for $a = P(z \neq z|x = x)$. By Proposition 5.1 and Corollary 4.3.2, sharp bounds on $E(y|x = x, z = z)$ are $e_{Lxz} \equiv \int y dL_{xz}$ and $e_{Uxz} \equiv \int y dU_{xz}$. Hence, $H[E(y|x = x, z)]$ is a convex subset of the hyper-rectangle $\times_{z \in Z} [e_{Lxz}, e_{Uxz}]$.

Third, observe that, by the Law of Iterated Expectations, $E(y|x = x, z)$ solves the linear equation

$$E(y|x = x) = \sum_{z \in Z} E(y|x = x, z)P(z = z|x = x). \qquad (5.7)$$

Hence $H[E(y|x = x, z)]$ is a convex set that lies in the intersection of the hyper-rectangle $\times_{z \in Z} [e_{Lxz}, e_{Uxz}]$ and the hyperplane that solves (5.7).

These immediate properties are useful but they do not completely describe $H[E(y|x = x, z)]$. Proposition 5.2, presented later in this section, goes much further by showing that $H[E(y|x = x, z)]$ has finitely many extreme points. These extreme points are the expectations of certain $|Z|$-tuples of *stacked distributions*, defined below. Corollary 5.2.1 shows that $H[E(y|x = x, z)]$ is the convex hull of its extreme points when Y has finite cardinality.

Stacked Distributions

Stacked distributions are sequences of $|Z|$ distributions such that the entire probability mass of the jth distribution lies weakly to the left of that of the (j+1)st distribution. To describe these distribution sequences, let Z now be the ordered set of integers $(1, \ldots, |Z|)$. This set has $|Z|!$ permutations, each of which generates a distinct $|Z|$-vector of stacked distributions. Label these $|Z|$-vectors $(P_{xj}^m, j = 1, \ldots, |Z|)$, $m = 1, \ldots, |Z|!$.

For each value of m, the elements of $(P_{xj}^m, j = 1, \ldots, |Z|)$ solve a recursive set of minimization problems. What follows shows the construction of $(P_{xj}^1, j = 1, \ldots, |Z|)$, which is based on the original ordering of Z. The other $(|Z|! - 1)$ vectors of distributions are generated in the same manner after permuting Z, which alters the order in which the recursion is performed.

For each $j = 1, \ldots, |Z|$, P_{xj}^1 is chosen to minimize its expectation subject to the distributions earlier chosen for $(P_{xi}^1, i < j)$ and subject to the global condition that equation (5.2) must hold. The recursion is as follows. For j $= 1, \ldots, |Z|$, P_{xj}^1 solves the problem

$$\min_{\psi \, \in \Gamma_Y} \int y d\psi \tag{5.8}$$

subject to

$$P(y|x = x) = \sum_{i=1}^{j-1} \pi_{xi} P_{xi}^1 + \pi_{xj} \psi + \sum_{k=j+1}^{|Z|} \pi_{xk} \psi_k, \tag{5.9}$$

where $\pi_{xj} \equiv P(z = j|x = x)$ and where $\psi_k \in \Gamma_Y$, $k = j + 1, \ldots, |Z|$ are unrestricted probability distributions.

This recursion yields a sequence of stacked distributions. For $j = 1$, equation (5.9) reduces to

$$P(y|x = x) = \pi_{x1} \psi + \sum_{k=2}^{|Z|} \pi_{xk} \psi_k. \tag{5.10}$$

The distribution solving (5.8) subject to (5.10) is L_{x1} defined in (5.6a), which is a right-truncated version of $P(y|x = x)$. Thus $P_{x1}^1 = L_{x1}$.

For $j = 2$, equation (5.9) has the form

$$P(y|x = x) = \pi_{x1}L_{x1} + \pi_{x2}\psi + \sum_{k=3}^{|Z|} \pi_{xk}\psi_k. \qquad (5.11)$$

Let υ_{x1} be the distribution that solves the equation

$$P(y|x = x) = \pi_{x1}L_{x1} + (1 - \pi_{x1})\upsilon_{x1}. \qquad (5.12)$$

Distribution υ_{x1} is a left-truncated version of $P(y|x = x)$ that has all of its mass to the right of L_{x1}. Combining (5.11) and (5.12) yields

$$\upsilon_{x1} = \frac{\pi_{x2}}{(1-\pi_{x1})}\psi + \sum_{k=3}^{|Z|} \frac{\pi_{xk}}{(1-\pi_{x1})}\psi_k. \qquad (5.13)$$

Equation (5.13) has the same form as (5.10), with υ_{x1} replacing $P(y|x = x)$ and $\pi_{x(k+1)}/(1 - \pi_{x1})$ replacing π_{xk}. Hence P_{x2}^1, the solution to (5.8) subject to (5.13), is a right-truncated version of υ_{x1}. Distributions P_{x1}^1 and P_{x2}^1 are stacked side-by-side, with all of the mass of the former distribution lying weakly to the left of the mass of the latter distribution. Distributions $\{P_{xj}^1, j = 3, \ldots, |Z|\}$ are similarly stacked. For each j, the mass of P_{xj}^1 lies weakly to the left of the mass of $P_{x(j+1)}^1$.

Stacking implies that, for each value of j, the supremum of the support of P_{x1j} may equal the infimum of the support of $P_{x(j+1)}^1$, but otherwise the distributions are concentrated on disjoint intervals. If $P(y|x = x)$ has a mass point, then P_{xj}^1 and $P_{x(j+1)}^1$ may share this mass point. However, if $P(y|x = x)$ is continuous, then P_{xj}^1 and $P_{x(j+1)}^1$ are continuous and place their mass on disjoint intervals.

The Extreme Points of the Identification Region

With the above as preliminary, Proposition 5.2 proves that the expectations of the stacked distributions are the extreme points of $H[E(y|x = x, z)]$. Then Corollary 5.2.1 shows that $H[E(y|x = x, z)]$ is the convex hull of these extreme points if Y has finite cardinality.

Proposition 5.2: Let $e_{mx} \equiv (\int y dP_{xj}^m, j = 1, \ldots, |Z|)$. The extreme points of $H[E(y|x = x, z)]$ are $\{e_{mx}, m = 1, \ldots, |Z|!\}$. $\qquad \square$

Proof: By construction, each vector in $\{e_{mx}, m = 1, \ldots, |Z|!\}$ is a feasible value of $E(y|x = x, z)$. Step (i) of the proof shows that these vectors are extreme points of $H[E(y|x = x, z)]$. Step (ii) shows that $H[E(y|x = x, z)]$ has no other extreme points. In this proof, the notation is simplified by everywhere suppressing the conditioning on the covariate x; for example, $E(y|x = x, z)$ and e_{mx} are abbreviated to $E(y|z)$ and e_m.

Step (i). It suffices to consider e_1. Permuting Z does not alter the argument.

Suppose that e_1 is not an extreme point of $H[E(y|z)]$. Then there exists an $\alpha \in (0, 1)$ and distinct vectors $(\xi', \xi'') \in H[E(y|z)]$ such that $e_1 = \alpha\xi' + (1 - \alpha)\xi''$. Suppose that e_1, ξ', and ξ'' differ in their first component. Then either $\xi_1' < e_{11} < \xi_1''$ or $\xi_1'' < e_{11} < \xi_1'$. By construction, $e_{11} = e_{L1}$, the global minimum of $E(y|z = 1)$. Hence, it must be the case that $\xi_1'' = \xi_1' = e_{11}$.

Now, suppose that e_1, ξ', and ξ'' differ in their second component. Then $\xi_2' < e_{12} < \xi_2''$ or $\xi_2'' < e_{12} < \xi_2'$. But e_{12} minimizes $E(y|z = 2)$ subject to the previous minimization of $E(y|z = 1)$. Hence $\xi_2' = \xi_2'' = e_{12}$. Recursive application of this reasoning shows that $\xi'' = \xi' = e_1$, contrary to supposition. Hence e_1 is an extreme point of $H[E(y|z)]$.

Step (ii). Let $\xi \in H[E(y|z)]$, with $\xi \notin \{e_m, m = 1, \ldots, |Z|!\}$. Then ξ is the expectation of some feasible $|Z|$-vector of non-stacked distributions. We want to show that ξ is not an extreme point of $H[E(y|z)]$. Thus, we must show that there exists an $\alpha \in (0, 1)$ and distinct $|Z|$-vectors $(\xi', \xi'') \in H[E(y|z)]$ such that $\xi = \alpha\xi' + (1 - \alpha)\xi''$.

Let the set-valued function $S(\psi)$ denote the support of any probability distribution ψ on the real line. Let $(\psi_j, j \in Z) \in (\Gamma_Y)^{|Z|}$ be any feasible $|Z|$-vector of distributions with expectation ξ. This $|Z|$-vector is not stacked, so there exist components ψ_i and ψ_k such that $[\inf S(\psi_i), \sup S(\psi_i)] \cap [\inf S(\psi_k), \sup S(\psi_k)]$ has positive length. Thus $\sup S(\psi_i) > \inf S(\psi_k)$ and $\sup S(\psi_k) > \inf S(\psi_i)$. For ease of exposition, henceforth let $a_j \equiv \inf S(\psi_j)$ and $b_j \equiv \sup S(\psi_j)$, for $j = i, k$.

Now, construct a feasible $|Z|$-vector of distributions that shifts mass, in a particular balanced manner, between distributions ψ_i and ψ_k, while leaving the other components of $(\psi_j, j \in Z)$ unchanged. Let $0 < \epsilon < \frac{1}{2}(b_i - a_k)$. Then $\psi_k[a_k, a_k+\epsilon] > 0$, $\psi_i[b_i-\epsilon, b_i] > 0$, and $[a_k, a_k+\epsilon] \cap [b_i-\epsilon, b_i] = \varnothing$. Let

$$\lambda \equiv \frac{\pi_k \, \psi_k[a_k, a_k + \epsilon]}{\pi_i \, \psi_i[b_i - \epsilon, b_i]}.$$

Now, define the new $|Z|$-vector $(\psi_j', j \in Z)$ as follows: Let $\psi_j' = \psi_j$ for $j \neq i, k$. If $\lambda \leq 1$, for $A \subset Y$ let

$[\psi_i'(A), \psi_k'(A)] =$

$$[\psi_i(A) + (\pi_k/\pi_i)\psi_k(A), \ 0] \qquad \text{if } A \subset [a_k, a_k + \epsilon]$$

$$[(1 - \lambda)\psi_i(A), \ \psi_k(A) + (\lambda\pi_i/\pi_k)\psi_i(A)] \ \text{if } A \subset [b_i - \epsilon, b_i]$$

$$[\psi_i(A), \ \psi_k(A)] \qquad\qquad\qquad\qquad \text{elsewhere.}$$

Alternatively, if $\lambda > 1$, let

$[\psi_i'(A), \psi_k'(A)] =$

$$[\psi_i(A) + (\pi_k/\lambda\pi_i)\psi_k(A), (1 - 1/\lambda)\psi_k(A)] \ \text{if } A \subset [a_k, a_k + \epsilon]$$

$$[0, \psi_k(A) + (\pi_i/\pi_k)\psi_i(A)] \qquad\qquad \text{if } A \subset [b_i - \epsilon, b_i]$$

$$[\psi_i(A), \ \psi_k(A)] \qquad\qquad\qquad\qquad \text{elsewhere.}$$

Thus, the new $|Z|$-vector shifts ψ_i mass leftward from the $[b_i - \epsilon, b_i]$ interval to the $[a_k, a_k + \epsilon]$ interval and compensates by shifting ψ_k mass rightward to the $[b_i - \epsilon, b_i]$ interval from the $[a_k, a_k + \epsilon]$ interval. The λ parameter ensures that we shift equal amounts of mass and that

$$\pi_i\psi_i' + \pi_k\psi_k' = \pi_i\psi_i + \pi_k\psi_k.$$

Hence $(\psi_j', j \in Z)$ is a feasible $|Z|$-vector of distributions. The mean of $(\psi_j', j \in Z)$ is related to the mean of $(\psi_j, j \in Z)$ as follows: $\xi_i' < \xi_i, \xi_k' > \xi_k$, and $\xi_j' = \xi_j$ for $j \neq i, k$.

An analogous operation switching the roles of i and k produces another $|Z|$-vector $(\psi_j'', j \in Z)$. Now let $0 < \epsilon < \frac{1}{2}(b_k - a_i)$ and redefine λ accordingly. This construction shifts ψ_k mass leftward from the $[b_k - \epsilon, b_k]$ interval to the $[a_i, a_i + \epsilon]$ interval and shifts an equal amount of ψ_i mass rightward to the $[b_k - \epsilon, b_k]$ interval from the $[a_i, a_i + \epsilon]$ interval, while ensuring that

$$\pi_i\psi_i'' + \pi_k\psi_k'' = \pi_i\psi_i + \pi_k\psi_k.$$

The mean of this $|Z|$-vector is related to the mean of $(\psi_j, j \in Z)$ as follows: $\xi_i'' > \xi_i, \xi_k'' < \xi_k$, and $\xi_j'' = \xi_j$ for $j \neq i, k$.

It follows from the above that

$$\pi_i\xi_i'' + \pi_k\xi_k'' = \pi_i\xi_i' + \pi_k\xi_k' = \pi_i\xi_i + \pi_k\xi_k.$$

Thus (ξ_i, ξ_k) lies on the line connecting (ξ_i', ξ_k') and (ξ_i'', ξ_k''). Moreover, $\xi_i'' > \xi_i > \xi_i'$ and $\xi_k' > \xi_k > \xi_k''$. Hence (ξ_i, ξ_k) is a strictly convex combination of

(ξ_i', ξ_k') and (ξ_i'', ξ_k''). Finally, recall that $\xi_j'' = \xi_j' = \xi_j$ for $j \neq i, k$. Hence ξ is a strictly convex combination of ξ' and ξ''. Thus ξ is not an extreme point of $H[E(y|z)]$.

Q. E. D.

Corollary 5.2.1: Let Y have finite cardinality $|Y|$. Then $H[E(y|x=x, z)]$ is the convex hull of its extreme points $\{e_{mx}, m = 1, \ldots, |Z|!\}$.   □

Proof: Minkowski's Theorem shows that a compact convex set in $R^{|Z|}$ is the convex hull of its extreme points.[1] We already know that $H[E(y|x=x, z)]$ is a bounded convex set, so we need only show that this set is closed. For $(y, j) \in Y \times Z$, let ϕ_{yj} be a feasible value for $P(y = y|x = x, z = j)$. Then equation (5.2) is this system of $|Y|$ linear equations in the $|Y| \times |Z|$ unknowns $\{\phi_{yj}, (y, j) \in Y \times Z\}$:

$$P(y = y|x = x) = \sum_{j \in Z} \pi_{xj}\phi_{yj}, \qquad y \in Y.$$

Let Φ denote the solutions to this system of equations. Φ forms a closed set in $R^{|Y| \times |Z|}$. The identification region for $E(y|x = x, z)$ is a linear map from Φ to $R^{|Z|}$, namely

$$H[E(y|x = x, z)] = \{(\sum_{y \in Y} y\phi_{yj}, j \in Z), \phi \in \Phi\}.$$

Hence $H[E(y|x = x, z)]$ is closed.

Q. E. D.

Corollary 5.2.1 completely describes $H[E(y|x = x, z)]$ when Y has finite cardinality. It is topologically delicate to determine if $H[E(y|x = x, z)]$ is closed when Y has infinite cardinality. This question is not addressed here.

5.4. Instrumental Variables

Propositions 5.1 and 5.2 characterize the restrictions on $E(y|x, z)$ implied by knowledge of $P(y|x)$ and $P(z|x)$, using the empirical evidence alone. Tighter inferences may be feasible if distributional assumptions are imposed.

Let us first dispose of an assumption whose implications are so immediate as barely to require comment. Suppose that y is known to be mean-independent of z, conditional on x, so $E(y|x, z) = E(y|x)$. Then knowledge of $P(y|x)$ per se point-identifies $E(y|x, z)$.[2]

This section examines two assumptions that use components of x as instrumental variables. Let $x = (v, w)$ and $X = V \times W$. One could assume that y is mean-independent of v, conditional on (w, z); that is,

$$E(y|x, z) = E(y|w, z). \tag{5.14}$$

Alternatively, one could assert that y is statistically independent of v, conditional on (w, z); that is,

$$P(y|x, z) = P(y|w, z). \tag{5.15}$$

Both assumptions use v as an instrumental variable, with assumption (5.15) being stronger than assumption (5.14).

Proposition 5.3 characterizes fully, albeit abstractly, the identifying power of assumptions (5.14) and (5.15). Then Corollary 5.3.1 presents a weaker, but much simpler, outer identification region that yields a straightforward *rank condition* for point identification of $E(y|w, z)$.

Proposition 5.3: Let $w \in W$. The identification regions for $E(y|w = w, z)$ under assumptions (5.14) and (5.15) are respectively

$$\mathbf{H}^*_w \equiv \bigcap_{v \in V} \mathbf{H}[E(y|v = v, w = w, z)] \tag{5.16}$$

and

$$\mathbf{H}^{**}_w = \{(\textstyle\int y d\eta_z, z \in Z), (\eta_z, z \in Z) \in \bigcap_{v \in V} \mathbf{H}[P(y|v = v, w = w, z)]\}. \tag{5.17}$$

The corresponding identification regions for $E(y|w, z)$ are $\times_{w \in W} \mathbf{H}^*_w$ and $\times_{w \in W} \mathbf{H}^{**}_w$. □

Proof: Consider assumption (5.14). For $(v, w) \in V \times W$, $(\eta_z, z \in Z) \in \mathbf{H}[P(y|v = v, w = w, z)]$ if and only if

$$P(y|v = v, w = w) = \sum_{z \in Z} \pi_{(v, w)z} \eta_z.$$

Let $\xi \in R^{|Z|}$. Under (5.14), ξ is a feasible value for $E(y|w = w, z)$ if and only if, for every $v \in V$, there exists an element of $\mathbf{H}[P(y|v = v, w = w, z)]$ having expectation ξ. $\mathbf{H}^*_w$ comprises these feasible values of $E(y|w = w, z)$.

Consider (5.15). Under this assumption, $(\eta_z, z \in Z)$ is a feasible value for $P(y|w = w, z)$ if and only if $(\eta_z, z \in Z)$ satisfies the system of equations

$$P(y|v = v, w = w) = \sum_{z \in Z} \pi_{(v, w)z} \eta_z, \qquad \forall v \in V.$$

Thus, the identification region for $[P(y|w = w, z), z \in Z]$ is

$$\cap_{v \in V} H[P(y|v = v, w = w, z)].$$

H_w^{**} comprises the expectations of these feasible vectors of distributions.

Now, consider $E(y|w, z)$. Neither (5.14) nor (5.15) imposes a cross-w restriction. Hence, the identification regions for $E(y|w, z)$ under these assumptions are the Cartesian products of the respective regions for $E(y|w = w, z)$, $w \in W$.

Q. E. D.

A Simple Outer Identification Region

Proposition 5.3 is too abstract to convey a sense of the identifying power of assumptions (5.14) and (5.15). Corollary 5.3.1 shows that a simple outer identification region emerges if one exploits only the Law of Iterated Expectations rather than the full force of the Law of Total Probability. The corollary also shows that assumption (5.14) is a refutable hypothesis.

Corollary 5.3.1: (a) Let assumption (5.14) hold. Let $w \in W$. Let $|V|$ denote the cardinality of V. Let Π denote the $|V| \times |Z|$ matrix whose zth column is $[\pi_{(v, w)z}, v \in V]$. Let $C_w^* \subset R^{|Z|}$ denote the set of solutions $\xi \in R^{|Z|}$ to the system of linear equations

$$E(y|v = v, w = w) = \sum_{z \in Z} \pi_{(v, w)z}\, \xi_z, \qquad \forall\, v \in V. \qquad (5.18)$$

Then $H_w^* \subset C_w^*$. If Π has rank $|Z|$, then C_w^* is a singleton and $H_w^* = C_w^*$.

(b) Let C_w^* be empty. Then assumption (5.14) does not hold.   □

Proof: (a) The Law of Iterated Expectations and assumption (5.14) imply that feasible values of $E(y|w = w, z)$ solve (5.18). Hence $H_w^* \subset C_w^*$. H_w^* is non-empty under assumption (5.14), so (5.18) must have at least one solution. If Π has rank $|Z|$, then (5.18) has a unique solution, implying that $H_w^* = C_w^*$.

(b) If C_w^* is empty, then H_w^* is empty. Hence (5.14) cannot hold.

Q. E. D.

Complement 5A. Structural Prediction

Social scientists often want to predict how an observed mean outcome $E(y)$ would change if the covariate distribution were to change from $P(x, z)$ to some other distribution, say $P^*(x, z)$. It is common to address this prediction problem under the assumption that the long mean regression $E(y|x, z)$ is *structural*, in the sense that this regression would remain invariant under the hypothesized change in the covariate distribution. Given this assumption, the mean outcome under covariate distribution $P^*(x, z)$ would be

$$E^*(y) \equiv \sum_{x \in X} \sum_{z \in Z} E(y|x = x, z = z) P^*(z = z|x = x)\, P^*(x = x).$$

To motivate the assumption that $E(y|x, z)$ is structural, social scientists sometimes pose behavioral models of the form $y = f(x, z, u)$, wherein a person's outcome y is some function f of the covariates (x, z) and of other factors u. Then $E(y|x, z)$ is structural if u is statistically independent of (x, z) and if the distribution of u remains unchanged under the hypothesized change in the covariate distribution.

What can be said about $E^*(y)$ when $E(y|x, z)$ is not identified? The findings of this chapter are applicable if the available data reveal $P(y|x)$ and $P(z|x)$. For example, a well-known problem in poverty research is to predict participation in social welfare programs under hypothesized changes in the geographic distribution and demographic attributes of the population. Let y indicate program participation, let x be a geographic unit such as a county, and let z denote demographic attributes. One may be willing to assume that $E(y|x, z)$ is structural in the sense defined above. Administrative records may reveal program participation by county, and census data may reveal demographic attributes by county; that is, $P(y|x)$ and $P(z|x)$.

In such settings, the findings of this chapter yield identification regions for $E(y|x, z)$ and hence for $E^*(y)$. For example, using the empirical evidence alone, one can conclude that $E^*(y)$ lies in the set

$$\left\{ \sum_{x \in X} \sum_{z \in Z} \xi_{xz}\, P^*(z = z|x = x) P^*(x = x);\ (\xi_{xz},\, z \in Z) \in H[E(y|x = x, z)],\, x \in X \right\}.$$

Endnotes

Sources and Historical Notes

The analysis in this chapter originally appeared in Cross and Manski (2002). In particular, Propositions 5.2 and 5.3 here are based on Propositions 1 and 3 there.

The early major contributions to analysis of the ecological inference problem appeared in sociology in the 1950s. Robinson (1950) criticized the then common practice of interpreting the *ecological correlation*, the cross-x correlation of $P(y|x)$ and $P(z|x)$, as the correlation of y and z. Soon afterwards two influential short papers were published in the same issue of the *American Sociological Review*. Considering settings in which y and z are binary random variables, Duncan and Davis (1953) and Goodman (1953) performed informal partial analyses of the identification problems that were eventually addressed in generality in Cross and Manski (2002). Duncan and Davis used numerical illustrations to demonstrate that knowledge of $P(y|x)$ and $P(z|x)$ implies a bound on $P(y|x, z)$. Goodman showed that point identification of $P(y|x, z)$ may be possible if the available data are combined with an assumption asserting that y is mean-independent of an instrumental variable. In this chapter, Proposition 5.1 formalizes the insight of Duncan and Davis, and Corollary 5.3.1 generalizes Goodman's finding.

The usage of the terms *short* and *long* in this chapter is borrowed from Goldberger (1991, Section 17.2), who calls $E(y|x)$ a *short regression* and $E(y|x, z)$ a *long regression*. A longstanding concern of the econometric literature on linear regression exposited by Goldberger has been to compare the parameter estimates obtained in a least squares fit of y to x with those obtained in a least squares fit of y to (x, z). The expected difference between the estimates obtained in the former and latter fits is sometimes called "omitted variable bias."

Comparison of short and long regressions has also been prominent in statistics. Stimulated by Simpson (1951), statisticians have been intrigued by the fact that the short regression $E(y|x)$ may be increasing in a scalar x and yet the long regression $E(y|x, z = z)$ may be decreasing in x for all $z \in Z$. Studies of *Simpson's Paradox* have sought to characterize the circumstances in which this phenomenon occurs (see, for example, Lindley and Novick, 1981; Zidek, 1984).

Text Notes

1. See Brøndsted (1983, Theorem 5.10).

2. Less obvious distributional assumptions that point-identify $P(y|x, z)$ have been studied by King (1997), who asserted achievement of "a solution to the ecological inference problem" in a book of that name. However, his assumptions immediately drew criticism, as evidenced in a dispute played out in the *Journal of the American Statistical Association* (Freedman, Klein, Ostland, and Roberts, 1998, 1999; King, 1999).

6

Response-Based Sampling

6.1. Reverse Regression

Consider once more the problem of prediction of outcomes conditional on covariates. As before, the random variable (y, x): $J \to Y \times X$ has distribution $P(y, x)$, and the objective is to learn the conditional distributions $P(y|x)$.

For $y \in Y$, let J_y denote the sub-population of persons who have outcome value y. Researchers sometimes observe covariate realizations drawn at random from the sub-populations J_y, $y \in Y$. This sampling process is known to epidemiologists studying the prevalence of disease as *case-control*, *case-referent*, or *retrospective* sampling. It is known to econometricians studying choice behavior as *choice-based* sampling or *response-based* sampling. The term *response-based* sampling will be used here.

Response-based sampling is often motivated by practical considerations. For example, epidemiologists have found that random sampling can be a costly way to gather data, so they have often turned to less expensive stratified sampling designs, especially response-based designs. One divides the population into ill ($y = 1$) and healthy ($y = 0$) response strata and samples at random within each stratum. Response-based designs are considered to be particularly cost-effective in generating observations of serious diseases, as ill persons are clustered in hospitals and other treatment centers.

Sampling at random from J_y, $y \in Y$ reveals the distributions $P(x|y)$ of covariates conditional on outcomes. The objective is to learn the distributions $P(y|x)$ of outcomes conditional on covariates, so response-based sampling poses this problem of inference from *reverse regression*: What does knowledge of $P(x|y)$ reveal about $P(y|x)$?

To simplify analysis, let the space $Y \times X$ be finite, with $P(x = x) > 0$, all $x \in X$. Then the identification problem is displayed by Bayes Theorem and the Law of Total Probability, which give

$$P(y = y \mid x = x) = \frac{P(x = x \mid y = y)P(y = y)}{P(x = x)}$$

$$= \frac{P(x = x \mid y = y)P(y = y)}{\sum_{y' \in Y} P(x = x \mid y = y')P(y = y')}, \quad (y, x) \in Y \times X.$$

(6.1)

Response-based sampling reveals $P(x \mid y)$ but is uninformative about the marginal outcome distribution $P(y)$. Hence the identification region for $P(y \mid x)$ using the empirical evidence alone is

$$H[P(y \mid x)] =$$

$$\left\{ \left[\frac{P(x = x \mid y = y)\gamma(y = y)}{\sum_{y' \in Y} P(x = x \mid y = y')\gamma(y = y')}, \quad (y, x) \in Y \times X \right], \quad \gamma \in \Gamma_Y \right\}.$$

(6.2)

Inspection of (6.2) shows that, for any given value $(y, x) \in Y \times X$, the empirical evidence is uninformative about $P(y = y \mid x = x)$; letting $\gamma(y = y)$ range over the interval $[0, 1]$ yields $H[P(y = y \mid x = x)] = [0, 1]$. Nevertheless, response-based sampling is informative about the manner in which $P(y = y \mid x = x)$ varies with x (see Section 6.4).

Econometricians and epidemiologists studying response-based sampling have point-identified features of $P(y \mid x)$ by combining the empirical evidence with various other forms of information. Section 6.2 describes the prevailing practice in econometrics, which has been to combine response-based sampling data with auxiliary data on the marginal distribution of outcomes or covariates. Section 6.3 describes the prevailing practice in epidemiology, which has focused attention on binary response settings (i.e., Y contains two elements) and has studied inference under the *rare-disease* assumption.

Sections 6.4 and 6.5 present findings on partial identification in binary response settings. Section 6.4 analyzes the identification region $H[P(y \mid x)]$

obtained using the empirical evidence alone and derives informative sharp bounds on the *relative risk* and *attributable risk* statistics commonly used in epidemiology. Section 6.5 examines inference when covariate data are observed for only one of the two sub-populations J_y, $y \in Y$.

6.2. Auxiliary Data on Outcomes or Covariates

Response-based sampling is problematic because the sampling process is uninformative regarding the marginal outcome distribution $P(y)$. The econometrics literature on response-based sampling has recommended solution of the problem by collection of auxiliary data that reveal $P(y)$.

Administrative records on population outcomes or an auxiliary survey of a random sample of the population may reveal $P(y)$ directly. Alternatively, administrative records on population covariates or an auxiliary random-sample survey may reveal the marginal covariate distribution $P(x)$. In the latter case, knowledge of $P(x)$ and $P(x|y)$ may be used to solve the Law of Total Probability

$$P(x = x) = \sum_{y \in Y} P(x = x | y = y) P(y = y), \qquad x \in X \quad (6.3)$$

for feasible values of $P(y)$.

Equation (6.3) generically has a unique solution if the cardinality of X is at least as large as that of Y. For example, let x and y be binary random variables, with $X = \{0, 1\}$ and $Y = \{0, 1\}$. Then (6.3) reduces to

$$P(x = 1) = P(x = 1 | y = 1)P(y = 1) + P(x = 1 | y = 0)[1 - P(y = 1)]. \quad (6.4)$$

Response-based sampling identifies $P(x = 1 | y = 1)$ and $P(x = 1 | y = 0)$. Hence, auxiliary data revealing $P(x)$ enable solution for $P(y)$, provided only that $P(x = 1 | y = 1) \neq P(x = 1 | y = 0)$.

6.3. The Rare-Disease Assumption

Epidemiologists often use response-based sampling to study the prevalence of diseases that are known to occur infrequently in the population. Let y be binary, with $y = 1$ if a person is ill with a specified disease and $y = 0$ otherwise. Epidemiologists use two statistics, relative risk and attributable risk, to measure how prevalence varies with observable covariates.

The *relative risk* (RR) of illness for persons with different covariate values, say $x = k$ and $x = j$, is

$$RR \equiv P(y=1 \mid x=k)/P(y=1 \mid x=j)$$

$$= \frac{P(x=k \mid y=1)}{P(x=j \mid y=1)} \cdot \frac{P(x=j \mid y=1)P(y=1) + P(x=j \mid y=0)P(y=0)}{P(x=k \mid y=1)P(y=1) + P(x=k \mid y=0)P(y=0)}.$$

$$(6.5)$$

The *attributable risk* (AR) is

$$AR \equiv P(y=1 \mid x=k) - P(y=1 \mid x=j)$$

$$= \frac{P(x=k \mid y=1)P(y=1)}{P(x=k \mid y=1)P(y=1) + P(x=k \mid y=0)P(y=0)}$$

$$- \frac{P(x=j \mid y=1)P(y=1)}{P(x=j \mid y=1)P(y=1) + P(x=j \mid y=0)P(y=0)}. \qquad (6.6)$$

In each case, the first identity defines the concept and the second equation follows from (6.1).

For example, let y indicate the occurrence of heart disease and let x indicate whether a person smokes cigarettes (yes = k, no = j). Then RR gives the ratio of the probability of heart disease conditional on smoking to the probability of heart disease conditional on not smoking, while AR gives the difference between these probabilities.

Texts on epidemiology discuss both relative and attributable risk, but empirical research has focused on relative risk. This focus is hard to justify from the perspective of public health. The health impact of altering a risk factor such as smoking presumably depends on the number of illnesses averted; that is, on the attributable risk times the size of the population. The relative risk statistic is uninformative about this quantity.

Yet relative risk continues to play a prominent role in epidemiological research. The rationale, such as it is, is that relative risk is point-identified under the *rare-disease* assumption, which lets the marginal probability of illness approach zero. As $P(y=1) \to 0$,

$$\lim_{P(y=1) \to 0} RR = \frac{P(x=k \mid y=1) \, P(x=j \mid y=0)}{P(x=j \mid y=1) \, P(x=k \mid y=0)} \qquad (6.7)$$

The sampling process reveals the quantities on the right side of (6.7), so the rare-disease assumption identifies relative risk. The expression on the right-side is called the *odds ratio* (OR) because it is the ratio of the odds of illness for persons with covariates k and j; that is, equation (6.1) yields

$$OR \equiv \frac{P(x = k \mid y = 1) \, P(x = j \mid y = 0)}{P(x = j \mid y = 1) \, P(x = k \mid y = 0)}$$

$$= \frac{P(y = 1 \mid x = k) \, P(y = 0 \mid x = j)}{P(y = 0 \mid x = k) \, P(y = 1 \mid x = j)}. \qquad (6.8)$$

Observe that the equality in (6.8) does not require the rare-disease assumption. The odds ratio is identified using the empirical evidence alone.

The rare-disease assumption also point-identifies attributable risk, but the result is unedifying. Letting $P(y = 1) \to 0$ yields

$$\lim_{P(y=1) \to 0} AR = 0. \qquad (6.9)$$

Thus, the rare-disease assumption implies that the disease is inconsequential from a public health perspective.

6.4. Bounds on Relative and Attributable Risk

This section shows what can be learned about relative and attributable risk from the empirical evidence alone, without using the rare-disease assumption or any other information. It is shown that response-based sampling partially identifies both quantities. As in Section 6.3, the present analysis supposes that y is binary.

Relative Risk

Examine the expression for relative risk in (6.5). Response-based sampling reveals all quantities on the right side except for $P(y)$. Hence the feasible values of RR may be determined by analyzing how the right side of (6.5) varies as $P(y = 1)$ ranges over the unit interval. The result, given in Proposition 6.1, is that relative risk must lie between the odds ratio and the value 1.

Proposition 6.1: Let $P(x|y = 1)$ and $P(x|y = 0)$ be known. Then the identification region for RR is

$$\text{OR} \leq 1 \quad \Rightarrow \quad \text{H(RR)} = [\text{OR}, 1], \qquad\qquad (6.10a)$$
$$\text{OR} \geq 1 \quad \Rightarrow \quad \text{H(RR)} = [1, \text{OR}]. \qquad\qquad (6.10b)$$

$\square$

Proof: Relative risk is a differentiable, monotone function of $P(y = 1)$, the direction of change depending on whether OR is less than or greater than one. To see this, let $p \equiv P(y = 1)$ and let $P_{im} \equiv P(x = i | y = m)$ for $i = j, k$, and $m = 0, 1$. Write RR explicitly as a function of p. Thus, define

$$RR_p \equiv \frac{P_{k1}}{P_{j1}} \cdot \frac{(P_{j1} - P_{j0})p + P_{j0}}{(P_{k1} - P_{k0})p + P_{k0}}.$$

The derivative of RR_p with respect to p is

$$\frac{P_{k1}}{P_{j1}} \cdot \frac{P_{j1}P_{k0} - P_{k1}P_{j0}}{[(P_{k1} - P_{k0})p + P_{k0}]^2}.$$

This derivative is positive if OR < 1, zero if OR $= 1$, and negative if OR $>$ 1. Hence, the extreme values of RR_p occur when p equals its extreme values of 0 and 1. Setting $p = 0$ makes RR $=$ OR, and setting $p = 1$ makes RR $= 1$. Intermediate values are feasible because RR_p is continuous in p.

Q. E. D.

Recall that the rare-disease assumption makes relative risk equal the odds ratio. Thus, this assumption always makes relative risks appear further from one than they actually are. The magnitude of the bias depends on the actual prevalence of the disease under study; the bias grows as $P(y = 1)$ moves away from zero.

Attributable Risk

Examine the expression for attributable risk in (6.6). Again, response-based sampling reveals all quantities on the right side except for $P(y)$. Hence the feasible values of AR may be determined by analyzing how the right side of (6.6) varies as $P(y = 1)$ ranges over the unit interval. Define

$$\text{ß} \equiv \left[\frac{P(x=j\,|\,y=1)P(x=j\,|\,y=0)}{P(x=k\,|\,y=1)P(x=k\,|\,y=0)} \right]^{\frac{1}{2}} \tag{6.11}$$

and

$$\pi \equiv$$

$$\frac{\text{ß}P(x=k\,|\,y=0) - P(x=j\,|\,y=0)}{[\text{ß}P(x=k\,|\,y=0) - P(x=j\,|\,y=0)] - [\text{ß}P(x=k\,|\,y=1) - P(x=j\,|\,y=1)]}. \tag{6.12}$$

Let

$$AR_\pi = \frac{P(x=k\,|\,y=1)\pi}{P(x=k\,|\,y=1)\pi + P(x=k\,|\,y=0)(1-\pi)}$$

$$- \frac{P(x=j\,|\,y=1)\pi}{P(x=j\,|\,y=1)\pi + P(x=j\,|\,y=0)(1-\pi)} \tag{6.13}$$

be the value that the attributable risk would take if $P(y=1) = \pi$. The result, given in Proposition 6.2, is that AR must lie between AR_π and zero.

Proposition 6.2: Let $P(x\,|\,y = 1)$ and $P(x\,|\,y = 0)$ be known. Then the identification region for AR is

$$OR \leq 1 \implies H(AR) = [AR_\pi, 0] \tag{6.14a}$$
$$OR \geq 1 \implies H(AR) = [0, AR_\pi]. \tag{6.14b}$$

$\square$

Proof: As $P(y = 1)$ increases from 0 to 1, the attributable risk changes parabolically, the orientation of the parabola depending on whether the odds ratio is less than or greater than one. To see this, again let $p \equiv P(y=1)$ and let $P_{im} \equiv P(x=i\,|\,y=m)$ for $i = j, k$, and $m = 0, 1$. Write AR explicitly as a function of p. Thus, define

$$AR_p \equiv \frac{P_{k1}p}{(P_{k1} - P_{k0})p + P_{k0}} - \frac{P_{j1}p}{(P_{j1} - P_{j0})p + P_{j0}}.$$

The derivative of AR_p with respect to p is

$$\frac{P_{k1}P_{k0}}{[(P_{k1} - P_{k0})p + P_{k0}]^2} - \frac{P_{j1}P_{j0}}{[(P_{j1} - P_{j0})p + P_{j0}]^2}.$$

The derivative equals zero at

$$\pi = \frac{\text{\ss}P_{k0} - P_{j0}}{(\text{\ss}P_{k0} - P_{j0}) - (\text{\ss}P_{k1} - P_{j1})}$$

and at

$$\pi^* = \frac{\text{\ss}P_{k0} + P_{j0}}{(\text{\ss}P_{k0} + P_{j0}) - (\text{\ss}P_{k1} + P_{j1})},$$

where $\text{\ss} \equiv (P_{j1}P_{j0}/P_{k1}P_{k0})^{1/2}$ was defined in (6.11). Examination of the two roots reveals that π always lies between zero and one, but π^* always lies outside the unit interval; so π is the only relevant root. Thus AR_p varies parabolically as p rises from zero to one.

Observe that $AR_p = 0$ at $p = 0$ and at $p = 1$. Examination of the derivative of AR_p at $p = 0$ and at $p = 1$ shows that the orientation of the parabola depends on the magnitude of the odds ratio. If OR < 1, then as p rises from zero to one, AR_p falls continuously from zero to its minimum at π and then rises back to zero. If OR > 1, then AR_p rises continuously from zero to its maximum at π and then falls back to zero. In the borderline case where OR = 1, AR_p does not vary with p.

Q. E. D.

6.5. Sampling from One Response Stratum

Methodological research on response-based sampling has concentrated on situations in which one samples at random from all of the sub-populations (J_y, $y \in Y$) and so learns all of the conditional covariate distributions $P(x|y)$. Often, however, one is able to sample from only a subset of these sub-populations. For example, an epidemiologist studying the prevalence of a disease may use hospital records to learn the distribution of covariates among persons who are ill ($y = 1$), but have no comparable data on persons who are healthy ($y = 0$). A social policy analyst studying participation in welfare programs may use the administrative records of the welfare system to learn the backgrounds of welfare recipients ($y = 1$), but have no comparable information on non-recipients ($y = 0$).

Suppose, as in these examples, that y is binary and that one can sample from sub-population J_1 but not from J_0, so response-based sampling reveals $P(x|y = 1)$ but not $P(x|y = 0)$. Inspection of (6.5) and (6.6) shows that knowledge of $P(x|y = 1)$ reveals nothing about relative and attributable risks. However, inference is possible if knowledge of $P(x|y=1)$ is combined with auxiliary data on the marginal outcome and/or covariate distribution.

Consider first the case in which auxiliary data collection reveals both marginal distributions, $P(y)$ and $P(x)$. Equation (6.1) shows that $P(y = 1|x)$ is point-identified. If y is binary, this means that $P(y|x)$ is point-identified.

Propositions 6.3 and 6.4 examine the cases in which one marginal distribution is known, but not the other. The propositions give identification regions for RR, AR, and the response probabilities themselves.

Proposition 6.3: Let $P(x|y = 1)$ and $P(y = 1)$ be known. Then the identification regions for RR and AR are

$$H(RR) = $$

$$\left[\frac{P(x = k|y = 1)P(y = 1)}{P(x = k|y = 1)P(y = 1) + P(y = 0)} , \frac{P(x = j|y = 1)P(y = 1) + P(y = 0)}{P(x = j|y = 1)P(y = 1)} \right],$$

$$(6.15)$$

$$H(AR) = $$

$$\left[-\frac{P(y = 0)}{P(x = k|y = 1)P(y = 1) + P(y = 0)} , \frac{P(y = 0)}{P(x = j|y = 1)P(y = 1) + P(y = 0)} \right].$$

$$(6.16)$$

For $x \in X$, the identification region for $P(y = 1|x = x)$ is

$$H[P(y = 1|x = x)] = \left[\frac{P(x = x|y = 1)P(y = 1)}{P(x = x|y = 1)P(y = 1) + P(y = 0)} , 1 \right]. \qquad (6.17)$$

□

Proof: Sharp lower (upper) bounds on RR and AR are obtained by letting $P(x = j | y = 0)$ equal zero (one) and $P(x = k | y = 0)$ equal one (zero) in (6.5) and (6.6). Intermediate values are feasible because RR and AR vary continuously with $P(x = j | y = 0)$ and $P(x = k | y = 0)$.

The sharp lower (upper) bound on $P(y = 1 | x = x)$ is obtained by letting $P(x = x | y = 0)$ equal one (zero) in (6.1). Intermediate values are feasible because $P(y = 1 | x = x)$ varies continuously with $P(x = x | y = 0)$.

Q. E. D.

Proposition 6.4: Let $P(x | y = 1)$ and $P(x)$ be known. Then RR is point-identified, with

$$RR = \frac{P(x = k | y = 1) \, P(x = j)}{P(x = j | y = 1) \, P(x = k)}. \tag{6.18}$$

Let $c \equiv \min_{i \in X} [P(x = i)/P(x = i | y = 1)]$. The identification region for $P(y = 1 | x = x)$ is the interval

$$H[P(y = 1 | x = x)] = [0, \, cP(x = x | y = 1)/P(x = x)]. \tag{6.19}$$

The identification region for AR is

$$d \leq 0 \; \Rightarrow \; H(AR) = [cd, 0] \tag{6.20a}$$
$$d \geq 0 \; \Rightarrow \; H(AR) = [0, cd], \tag{6.20b}$$

where $d \equiv [P(x = k | y = 1)/P(x = k) - P(x = j | y = 1)/P(x = j)]$. □

Proof: Recall equation (6.1), whose first equality (Bayes Theorem) gives

$$P(y = 1 | x = x) = \frac{P(x = x | y = 1)P(y = 1)}{P(x = x)}, \quad x \in X.$$

Equation (6.18) follows from this and the definition of relative risk.

Now, fix x and consider inference on $P(y = 1 | x = x)$. The quantities $P(x = x | y = 1)$ and $P(x = x)$ are known by assumption. $P(y = 1)$ is not known but must lie in the interval $[0, c]$. To show this, let i be any element of X, and write $P(x = i)$ as

$$P(x = i) = P(x = i | y = 1)P(y = 1) + P(x = i | y = 0)[1 - P(y = 1)].$$

Solve for $P(x = i \mid y = 0)$, yielding

$$P(x = i \mid y = 0) = \frac{P(x = i) - P(x = i \mid y = 1)P(y = 1)}{1 - P(y = 1)}.$$

The probabilities $\{P(x = i \mid y = 0), i \in X\}$ must satisfy the inequalities $\{0 \le P(x = i \mid y = 0), i \in X\}$ and must sum to one. These probabilities sum to one for all values of $P(y = 1)$, but the inequalities $\{0 \le P(x = i \mid y = 0), i \in X\}$ hold if and only if $P(y = 1) \le c$. This yields (6.19).

Finally, consider attributable risk. By definition of AR and d, AR = $P(y = 1)d$. The empirical evidence reveals d, and it has been found that $P(y = 1) \in [0, c]$. Hence (6.20) holds.

Q. E. D.

Complement 6A. Smoking and Heart Disease

A numerical example concerning smoking and heart disease illustrates the findings. This example is drawn from Manski (1995, Chapter 4).

Let y indicate the occurrence of heart disease. Let the probabilities of heart disease conditional on smoking ($x = k$) and nonsmoking ($x = j$) be 0.12 and 0.08. Let the fraction of persons who smoke be 0.50. These values imply that the marginal probability of heart disease is 0.10, and that the probabilities of smoking conditional on being ill and healthy are 0.60 and 0.49. The implied odds ratio is 1.57, relative risk is 1.50, and attributable risk is 0.04. Thus, the parameters of the example are

$P(y = 1 \mid x = k) = 0.12$ $P(y = 1 \mid x = j) = 0.08$ $P(x = k) = P(x = j) = 0.50$
$P(y = 1) = 0.10$ $P(x = k \mid y = 1) = 0.60$ $P(x = k \mid y = 0) = 0.49$
$OR = 1.57$ $RR = 1.50$ $AR = 0.04$.

The researcher does not know RR and AR. However, Proposition 6.1 shows that RR lies between OR and one, so the empirical evidence reveals that RR $\in [1, 1.57]$. The quantities (β, π, AR_π) defined in Proposition 6.2 take the values $\beta = 0.83$, $\pi = 0.51$, and $AR_\pi = 0.11$. Hence, the empirical evidence reveals that AR $\in [0, 0.11]$.

Using Prior Information to Tighten the Bounds

The calculations above assume no prior information about the marginal probability $P(y = 1)$. Someone possessing information about the prevalence of heart disease in the population can tighten the bounds. A simple way to do this is to extend Propositions 6.1 and 6.2 to cases in which $P(y = 1)$ is permitted to vary over a restricted range rather than over the interval [0, 1].

King and Zeng (2002) extend the propositions in this manner and reconsider the numerical example above when $P(y = 1)$ is permitted to vary over the range [0.05, 0.15]. Under the assumption that $P(y = 1)$ lies in this interval, they find that RR $\in$ [1.46, 1.53] and AR $\in$ [0.021, 0.056].

Sampling from One Response Stratum

Suppose that one has a random sample of ill persons but not of healthy ones, so $P(x \mid y = 1)$ is known but not $P(x \mid y = 0)$. Consider Propositions 6.3 and 6.4 in this setting.

First, suppose that the outcome distribution is known. By Proposition 6.3, $P(y = 1 \mid x = k) \in$ [0.06, 1], $P(y = 1 \mid x = j) \in$ [0.04, 1], RR $\in$ [0.06, 23.5], and AR $\in$ [-0.94, 0.96]. Thus, knowing the marginal probability of heart disease has little identifying power in this example.

Now, suppose that the covariate distribution is known. The parameters of the example give c = 0.83, so Proposition 6.4 yields $P(y = 1 \mid x = k) \in$ [0, 1] and $P(y = 1 \mid x = j) \in$ [0, 0.67]. The quantity d = 0.4, so AR $\in$ [0, 0.33]. Thus, knowing the prevalence of smoking in the population reveals little about the magnitudes of the response probabilities but reveals a fair bit about attributable risk and point-identifies relative risk.

Endnotes

Sources and Historical Notes

The analysis in Sections 6.4 and 6.5 originally appeared in Manski (1995, 2001). In particular, Propositions 6.1 and 6.2 are based on Manski (1995, Chapter 4), and Propositions 6.3 and 6.4 on Manski (2001, Propositions 3 and 4). Manski and Lerman (1977) recommended collection of auxiliary outcome data to learn $P(y)$. Hsieh, Manski, and McFadden (1985) showed that auxiliary covariate data can enable deduction of $P(y)$.

Cornfield (1951) showed that the rare-disease assumption point-identifies relative risk. The lack of relevance of relative risk to public health has long been criticized; see, for example, Berkson (1958), Fleiss (1981, Section 6.3), and Hsieh, Manski, and McFadden (1985).

The term *reverse regression* has been used in the literature on mean regression to denote the conditional expectation $E(x \mid y)$, when the conditional expectation of interest is $E(y \mid x)$; see Goldberger (1984).

7

Analysis of Treatment Response

7.1. Anatomy of the Problem

The four remaining chapters of this book study a pervasive and distinctive problem of missing outcomes. The problem is the non-observability of *counterfactual outcomes* in empirical analysis of treatment response.

Studies of treatment response aim to predict the outcomes that would occur if different treatment rules were applied to a population. Treatments are mutually exclusive, so one cannot observe the outcomes that a person would experience under all treatments. At most, one can observe the outcome that a person experiences under the treatment actually received. The counterfactual outcomes that a person would have experienced under other treatments are logically unobservable.

For example, suppose that patients ill with a specified disease can be treated by drugs or by surgery. The relevant outcome might be life span. One may want to predict the life spans that would occur if patients with specified risk factors were to receive each treatment. The available data may be observations of the actual life spans of these patients, some of whom were treated by drugs and the rest by surgery.

Predicting Outcomes Under Conjectural Treatment Rules

To formalize the inferential problem, let each member j of population J have covariates $x_j \in X$ and a *response function* $y_j(\cdot)$: $T \rightarrow Y$ mapping the mutually exclusive and exhaustive treatments $t \in T$ into outcomes $y_j(t) \in Y$. Let $z_j \in T$ denote the treatment that person j receives and $y_j \equiv y_j(z_j)$ be the outcome that he experiences. Then $y_j(t), t \neq z_j$ are counterfactual outcomes.

99

Let $y(\cdot): J \to Y^{|T|}$ be the random variable mapping the population into their response functions. Let $z: J \to T$ be the *status quo treatment rule* mapping the members of J into the treatments that they actually receive. Response functions are not observable, but covariates, realized treatments, and realized outcomes may be observable. If so, random sampling from J reveals the status quo (outcome, treatment) distributions $P(y, z|x)$ as well as the covariate distribution $P(x)$.

The distinctive problem of the analysis of treatment response is to predict the outcomes that would occur under alternatives to the status quo treatment rule. Let $\tau: J \to T$ be a *conjectural treatment rule*, the outcomes of which one would like to predict. Thus, person j's outcome under rule τ would be $y_j(\tau_j)$. This outcome is counterfactual whenever $\tau_j \neq z_j$. Hence, the sampling process does not reveal the conjectural outcome distributions $P[y(\tau)|x]$. The problem is to combine empirical knowledge of $P(y, z|x)$ with credible prior information to learn about $P[y(\tau)|x]$.

To simplify the presentation, the analysis in this chapter supposes that the covariate space X is finite and that $P(x = x, z = t) > 0$, $(t, x) \in T \times X$. These regularity conditions are maintained without further reference.

The Selection Problem

Researchers studying treatment response often want to predict the outcomes that would occur under conjectural treatment rules in which all persons with the same covariates receive the same treatment. Consider, for example, the medical setting described earlier. Let the relevant covariate be age. Then one treatment rule might mandate that all patients receive drugs, another that all patients receive surgery, and yet another that older patients receive drugs and younger ones receive surgery.

By definition, $P[y(t)|x = x]$ is the distribution of outcomes that would occur if all persons with covariate value x were to receive a specified treatment t. Hence prediction of outcomes under rules mandating uniform treatment conditional on covariates requires inference on the distributions $\{P[y(t)|x], t \in T\}$. The problem of identification of these distributions from knowledge of $P(y, z|x)$ is commonly called the *selection problem*.

The selection problem has the same structure as the missing-outcomes problem of Chapter 1 and Section 3.2. To see this, write

$$P[y(t)|x = x]$$

$$= P[y(t)|x = x, z = t]P(z = t|x = x) + P[y(t)|x = x, z \neq t]P(z \neq t|x = x)$$

$$= P(y|x = x, z = t)P(z = t|x = x) + P[y(t)|x = x, z \neq t]P(z \neq t|x = x). \quad (7.1)$$

The first equality is the Law of Total Probability. The second holds because $y(t)$ is the outcome experienced by persons who receive treatment t. The sampling process reveals $P(y|x = x, z = t)$, $P(z = t|x = x)$, and $P(z \neq t|x = x)$, but it is uninformative about $P[y(t)|x = x, z \neq t]$. Hence, the identification region for $P[y(t)|x = x]$ using the empirical evidence alone is

$$H\{P[y(t)|x = x]\} =$$

$$\{P(y|x = x, z = t)P(z = t|x = x) + \gamma P(z \neq t|x = x), \gamma \in \Gamma_Y\}. \quad (7.2)$$

Now, consider the collection of conjectural outcome distributions $\{P[y(t)|x], t \in T\}$. The sampling process is jointly uninformative about the counterfactual outcome distributions $\{P[y(t)|x = x, z \neq t], t \in T, x \in X\}$, which can take any value in $\times_{(t, x) \in T \times X} \Gamma_Y$. Hence the identification region for $\{P[y(t)|x], t \in T\}$ using the empirical evidence alone is the Cartesian product

$$H\{P[y(t)|x], t \in T\} = \times_{(t, x) \in T \times X} H\{P[y(t)|x = x]\}. \quad (7.3)$$

Observe that the empirical evidence cannot refute the hypothesis that all treatments have the same outcome distribution, conditional on x. Consider the hypothesis: $P[y(t)|x] = P[y(t')|x]$, all $(t, t') \in T \times T$. Identification region (7.3) necessarily contains distributions that satisfy this hypothesis. The easiest way to see this is to observe that the empirical evidence cannot refute the much stronger hypothesis $\{y_j(t) = y_j, \text{ all } (t, j) \in T \times J\}$; that is, each person's counterfactual outcomes could in principle be the same as the outcome that he actually experiences.

Random Treatment Selection

A familiar "solution" to the selection problem is to assume that the status quo treatment rule makes realized treatments statistically independent of response functions, conditional on x; that is,

$$P[y(\cdot)|x] = P[y(\cdot)|x, z]. \quad (7.4)$$

This assumption implies that, for each $t \in T$ and $x \in X$,

$$P[y(t)|x = x] = P[y(t)|x = x, z = t] = P(y|x = x, z = t). \quad (7.5)$$

The sampling process reveals $P(y|x = x, z = t)$. Hence, assumption (7.4) point-identifies $P[y(t)|x = x]$.

Assumption 7.4 is credible in a classical randomized experiment, in which an explicit randomization mechanism has been used to assign treatments and all persons comply with their treatment assignments. The credibility of the assumption in other applied settings almost invariably is a matter of controversy.

The Task Ahead

This opening section introduced the general problem of predicting outcomes under conjectural treatment rules and then examined basic elements of the selection problem. The remainder of this chapter uses a social-planning problem to motivate rules mandating uniform treatment conditional on covariates and to show how the selection problem affects treatment choice. Chapters 8 and 9 study the identifying power of some monotonicity assumptions that may be credible and useful in the analysis of treatment response. Chapter 10 studies the *mixing problem*; that is, the problem of predicting outcomes under conjectural treatment rules that do not mandate uniform treatment conditional on covariates.

7.2. Treatment Choice in Heterogeneous Populations

An important practical objective of empirical studies of treatment response is to provide decision makers with information useful in choosing treatments. Often the decision maker is a *planner* who must choose treatments for a heterogeneous *treatment population*. The planner may want to choose treatments whose outcomes maximize the welfare of the treatment population.

For example, consider a physician choosing medical treatments for a population of patients. The physician may observe each patient's demographic attributes, medical history, and the results of diagnostic tests. He may then choose a treatment rule that makes treatment a function of these covariates. If the physician acts on behalf of his patients, the outcome of interest may be a measure of patient health status and welfare may be this measure of health status minus the cost of treatment, measured in comparable units.

As another example, consider a judge choosing sentences for a population of convicted offenders. The judge may observe each offender's past criminal record, demeanor in court, and other attributes. Subject to legislated sentencing guidelines, she may consider these covariates when choosing sentences. If the judge acts on behalf of society, the outcome of interest may be a measure of recidivism, and welfare may be this measure of recidivism minus the cost of carrying out a sentence.

I present here a simple formulation of the planner's problem that motivates inference on the outcome distributions $\{P[y(t)|x], t \in T\}$ in general and the conditional mean outcomes $\{E[y(t)|x], t \in T\}$ in particular. I first specify the planner's choice set and objective function. I then derive the optimal treatment rule.

The Choice Set

Suppose that there is a finite set T of mutually exclusive and exhaustive treatments. A planner must choose a treatment for each member of the treatment population, denoted J^*. Each member j of population J^* has observable covariates $x_j \in X$ and an unobservable response function $y_j(\cdot)$: $T \rightarrow Y$ mapping treatments into real-valued outcomes.

The treatment population J^* is *identical in distribution* to the study population J, in which treatments have already been selected and outcomes have been realized. Thus (J^*, Ω, P) is a probability space whose probability measure P coincides with that of (J, Ω, P). The only difference between J and J^* is that the status quo treatment rule z has been applied in the former population, whereas a treatment rule has yet to be chosen in the latter.

There are no budgetary or other constraints that make it infeasible to choose some treatment rules. However, the planner cannot distinguish among persons with the same observed covariates and so cannot implement treatment rules that systematically differentiate among these persons. Hence, the feasible non-randomized rules are functions mapping the observed covariates into treatments.[1] Thus, uniform treatment conditional on covariates emerges naturally out of the planner's problem.

Formally, let Z(X) be the space of all functions mapping X into T. Let $z(\cdot) \in Z(X)$. Then feasible treatment rules have the form $\tau_j = z(x_j)$, $j \in J^*$.

The Objective Function

Suppose that the planner wants to maximize population mean welfare. Let the welfare obtained from assigning treatment t to person j have the additive form $y_j(t) + c(t, x_j)$. Here $c(\cdot, \cdot): T \times X \rightarrow R$ is a real-valued *cost function* known to the planner at the time of treatment choice. For each $z(\cdot) \in Z(X)$, let $E\{y[z(x)] + c[z(x), x]\}$ denote the mean welfare that would occur if the planner were to choose treatment rule $z(\cdot)$. Then the planner wants to solve the problem

$$\max_{z(\cdot) \in Z(X)} \quad E\{y[z(x)] + c[z(x), x]\}. \tag{7.6}$$

In the case of a physician, for example, $y_j(t)$ may measure the health status of patient j following receipt of treatment t, and $c(t, x_j)$ may be the (negative-valued) cost of treatment. The physician may know the costs of alternative treatments but not their health outcomes. Similarly, in the case of a judge, $y_j(t)$ may measure the rate of recidivism of offender j following receipt of sentence t, and $c(t, x_j)$ may be the cost of carrying out the sentence. Again, the judge may know the costs of alternative sentences but not their recidivism outcomes.

The assumption that the planner wants to maximize population mean welfare has normative, analytical, and practical appeal. This criterion function is standard in the public economics literature on social planning, which assumes that the objective is to maximize a utilitarian social welfare function. Linearity of the expectation operator yields substantial analytical simplifications, particularly through use of the law of iterated expectations. The practical appeal is that, as shown below, a planner choosing treatments to maximize mean welfare will want to learn mean treatment response, the main statistic reported in empirical studies of treatment response.

The Optimal Treatment Rule

The solution to optimization problem (7.6) is to assign to each member of the population a treatment that maximizes mean welfare conditional on the person's observed covariates. To show this, let $1[\cdot]$ be the indicator function taking the value one if the logical condition in the brackets holds and the value zero otherwise. For each $z(\cdot) \in Z(X)$, use the Law of Iterated Expectations to write

$$E\{y[z(x)] + c[z(x), x]\} = E\{E\{y[z(x)] + c[z(x), x] \mid x\}\}$$

$$= \sum_{x \in X} P(x = x)\{\sum_{t \in T} \{E[y(t) \mid x] + c(t, x)\} \cdot 1[z(x) = t]\}. \quad (7.7)$$

For each $x \in X$, the quantity $\sum_{t \in T} \{E[y(t) \mid x = x] + c(t, x)\} \cdot 1[z(x) = t]$ is maximized by choosing $z(x)$ to maximize $E[y(t) \mid x] + c(t, x)$ on $t \in T$. Hence, a treatment rule $z^*(\cdot)$ is optimal if , for each $x \in X$, $z^*(x)$ solves the problem

$$\max_{t \in T} E[y(t) \mid x = x] + c(t, x). \quad (7.8)$$

A planner who knows the conditional mean outcomes $E[y(\cdot) \mid x] \equiv \{E[y(t) \mid x = x], t \in T, x \in X\}$ can implement the optimal treatment rule. However, the selection problem and other identification problems limit the information that studies of treatment response provide. Sections 7.3 and 7.4

show how the selection problem affects treatment choice. Complements 7A and 7C consider the implications of other identification problems.

7.3. The Selection Problem and Treatment Choice

Let Y contain its lower and upper bounds $y_0 \equiv \inf_{y \in Y}$ and $y_1 \equiv \sup_{y \in Y}$. For $t \in T$ and $x \in X$, use the Law of Iterated Expectations to write

$E[y(t) \mid x = x] =$

$E(y \mid x = x, z = t) \cdot P(z = t \mid x = x) + E[y(t) \mid x = x, z \neq t] \cdot P(z \neq t \mid x = x).$ (7.9)

The empirical evidence reveals $E(y \mid x = x, z = t)$ and $P(z \mid x = x)$, but it is uninformative about $E[y(t) \mid x = x, z \neq t]$. Hence, the identification region for $E[y(t) \mid x = x]$ using the empirical evidence alone is the closed interval

$H\{E[y(t) \mid x = x]\}$

$$= [E(y \mid x = x, z = t) \cdot P(z = t \mid x = x) + y_0 \cdot P(z \neq t \mid x = x),$$
$$E(y \mid x = x, z = t) \cdot P(z = t \mid x = x) + y_1 \cdot P(z \neq t \mid x = x)]. \quad (7.10)$$

This result is a direct application of Proposition 1.1.

The object of interest is the collection of conditional mean outcomes $E[y(\cdot) \mid x]$. Its identification region $H\{E[y(\cdot) \mid x]\}$ is the rectangle

$H\{E[y(\cdot) \mid x]\}$

$$= \times_{(t, x) \in T \times X} [E(y \mid x = x, z = t) \cdot P(z = t \mid x = x) + y_0 \cdot P(z \neq t \mid x = x),$$
$$E(y \mid x = x, z = t) \cdot P(z = t \mid x = x) + y_1 \cdot P(z \neq t \mid x = x)]. \quad (7.11)$$

This is the identification region because the empirical evidence is uninformative about the set of counterfactual means $\{E[y(t) \mid x = x, z \neq t], (t, x) \in T \times X\}$.

$H\{E[y(\cdot) \mid x]\}$ is a bounded, proper subset of $(Y^{|T|} \times X)$ if Y has bounded range. Suppose that this is so. Then, without loss of generality, outcomes may be scaled to take values in the unit interval. Setting $y_0 = 0$ and $y_1 = 1$ gives $H\{E[y(\cdot) \mid x]\}$ the simpler form

$$H\{E[y(\cdot)|x]\} = \times_{(t,\,x)\,\in\,T\,\times\,X} [E(y|x=x,\,z=t)\cdot P(z=t|x=x),$$
$$E(y|x=x,\,z=t)\cdot P(z=t|x=x) + P(z\neq t|x=x)].$$
$$(7.12)$$

The analysis in the remainder of this chapter assumes that Y has bounded range and that outcomes have been scaled to lie in the unit interval. Thus (7.12) henceforth gives the identification region for $E[y(\cdot)|x]$ using the empirical evidence alone.

Dominated Treatment Rules

In general, the empirical evidence alone is not sufficiently informative about $E[y(\cdot)|x]$ to enable solution of optimization problem (7.8). What should the planner do?

Clearly, the planner should not choose a *dominated* treatment rule. (A treatment rule $z(\cdot)$ is dominated if there exists another feasible rule, say $z'(\cdot)$, that necessarily yields at least the mean welfare of $z(\cdot)$ and that performs strictly better than $z(\cdot)$ in some possible state of nature.) The rectangular form of $H\{E[y(\cdot)|x]\}$ makes it easy to determine what treatment rules are dominated.

Let $(t, x) \in T \times X$, and consider any rule that assigns treatment t to persons with covariate value x. By (7.12), the mean welfare yielded by this treatment choice can take any value in the interval

$$[E(y|x=x,\,z=t)\cdot P(z=t|x=x) + c(t,\,x),$$
$$E(y|x=x,\,z=t)\cdot P(z=t|x=x) + P(z\neq t|x=x) + c(t,\,x)].$$

The mean welfare of another treatment choice, say t', can take any value in the interval

$$[E(y|x=x,\,z=t')\cdot P(z=t'|x=x) + c(t',\,x),$$
$$E(y|x=x,\,z=t')\cdot P(z=t'|x=x) + P(z\neq t'|x=x) + c(t',\,x)].$$

Treatment t is definitely inferior to t' if the upper bound of the former interval is no larger than the lower bound of the latter one. This gives Proposition 7.1.

Proposition 7.1: Let $(t, x) \in T \times X$. Using the empirical evidence alone, a treatment rule that assigns treatment t to persons with covariates x is dominated if and only if there exists a treatment $t' \in T$ such that

$$E(y \mid x = x, z = t) \cdot P(z = t \mid x = x) + P(z \neq t \mid x = x) + c(t, x)$$

$$\leq \quad E(y \mid x = x, z = t') \cdot P(z = t' \mid x = x) + c(t', x). \qquad (7.13)$$

$$\square$$

The special case in which all treatments have the same cost is noteworthy. Then there generically are no dominated treatments. To see this, let $c(t, x) = c(t', x)$ for all treatments t and t'. Then inequality (7.13) reduces to

$$E(y \mid x = x, z = t) \cdot P(z = t \mid x = x) + P(z \neq t \mid x = x)$$

$$\leq \quad E(y \mid x = x, z = t') \cdot P(z = t' \mid x = x).$$

Observe that $P(z \neq t \mid x = x) \geq P(z = t' \mid x = x)$, $E(y \mid x = x, z = t) \in [0, 1]$, and $E(y \mid x = x, z = t') \in [0, 1]$. Hence, this inequality can never hold strictly. Moreover, it holds weakly only when $P(z \neq t \mid x = x) = P(z = t' \mid x = x)$, $E(y \mid x = x, z = t) = 0$, and $E(y \mid x = x, z = t') = 1$.

Choice Among Undominated Treatments

The fact that the empirical evidence does not enable determination of the optimal treatment rule does not imply that a planner should be paralyzed, unwilling and unable to choose a rule. It implies only that the planner cannot assert optimality for whatever rule he does choose.

The planner could, for example, reasonably apply the maximin rule, which calls for persons with covariates x to be assigned the treatment that maximizes the lower bound on $E[y(\cdot) \mid x = x]$. By (7.12), the maximin rule solves the problem

$$\max_{t \in T} E(y \mid x = x, z = t) \cdot P(z = t \mid x = x) + c(t, x). \qquad (7.14)$$

This rule is simple to apply and to comprehend. From the maximin perspective, the desirability of treatment t increases with $E(y \mid x = x, z = t)$, the mean outcome experienced by persons who received this treatment, and with $P(z = t \mid x = x)$, the fraction of persons who received treatment t. The second factor gives form to the conservatism of the maximin rule—the more prevalent a treatment was in the study population, the more expedient it is to choose this treatment in the treatment population.

7.4. Instrumental Variables

Section 7.3 considered a planner who confronts the selection problem using the empirical evidence alone. Credible distributional assumptions may enable the planner to shrink the identification region for $E[y(\cdot)|x]$ and hence shrink the set of undominated treatment rules. Many distributional assumptions make use of instrumental variables.

Treatment-Specific Assumptions

The selection problem is a matter of missing outcome data, so all of the analysis of Chapter 2 may be brought to bear. Thus, suppose that person j is characterized by an observable covariate v_j in a finite space V. Let $P(y, z, x, v)$ denote the joint distribution of (y, z, x, v). The covariate v serving as an instrumental variable need not be distinct from the covariate x used to make treatment choices, but it simplifies analysis if v contains information not conveyed by x. Hence, the presentation here assumes that $P(v = v, z = t|x) > 0$ for all $(v, t) \in V \times T$.

Let $t \in T$ and $x \in X$. To help identify $E[y(t)|x = x]$, a planner could impose any of the distributional assumptions studied in Chapter 2. This section shows how Assumptions MAR, SI, MMAR, and MI may be applied to the analysis of treatment response. Assumptions MM and MMM will be considered separately in Chapter 9.

In the context of treatment response, Assumptions MAR, SI, MMAR, and MI are as follows.

Outcomes Missing-at-Random (Assumption MAR):

$$P[y(t)|x = x, v] = P[y(t)|x = x, v, z = t] = P[y(t)|x = x, v, z \neq t]. \quad (7.15)$$

Statistical Independence of Outcomes and Instruments (Assumption SI):

$$P[y(t)|x = x, v] = P[y(t)|x = x]. \quad (7.16)$$

Means Missing-at-Random (Assumption MMAR):

$$E[y(t)|x = x, v] = E[y(t)|x = x, v, z = t] = E[y(t)|x = x, v, z \neq t]. \quad (7.17)$$

Mean Independence of Outcomes and Instruments (Assumption MI):

$$E[y(t)|x = x, v] = E[y(t)|x = x]. \quad (7.18)$$

These assumptions restrict the distribution of outcomes under a specified treatment t for persons with specified covariates x. A planner can consider each value of (t, x) in turn and decide what assumption to assert.

Assumptions MAR and MMAR point-identify $E[y(t)|x = x]$. The other assumptions generally do not yield point identification but do shrink the identification region. Propositions 7.2 through 7.5 give the results. These propositions are immediate extensions of corresponding ones in Chapter 2, so proofs are omitted.

Proposition 7.2: Let assumption MAR hold. Then $P[y(t)|x = x]$ is point-identified with

$$P[y(t)|x = x] = \sum_{v \in V} P(y|x = x, v = v, z = t)P(v = v \,|x = x). \quad (7.19)$$
$$\square$$

Proposition 7.3: Let assumption SI hold. Then the identification region for $P[y(t)|x = x]$ is

$$H_{SI}\{P[y(t)|x = x]\} = \bigcap_{v \in V} \{P(y|x = x, v = v, z = t)P(z = t|x = x, v = v)$$
$$+ \gamma_v \cdot P(z \neq t|x = x, v = v), \ \gamma_v \in \Gamma_Y\}. \quad (7.20)$$
$$\square$$

Proposition 7.4: Let assumption MMAR hold. Then $E[y(t)|x = x]$ is point-identified with

$$E[y(t)|x = x] = \sum_{v \in V} E(y|x = x, v = v, z = t)P(v = v|x = x). \quad (7.21)$$
$$\square$$

Proposition 7.5: Let assumption MI hold. Then the identification region for $E[y(t)|x = x]$ is the closed interval

$$H_{MI}\{E[y(t)|x = x]\} = [\max_{v \in V} E\{y \cdot 1[z = t]|x = x, v = v\},$$
$$\min_{v \in V} E\{y \cdot 1[z = t] + 1[z \neq t]|x = x, v = v\}]. \quad (7.22)$$
$$\square$$

Statistical Independence of Response Functions and Instruments

Whereas the assumptions considered above are treatment-specific, one could instead have information that restricts the joint distribution of the response function $y(\cdot)$. An especially prominent assumption is

Statistical Independence of Response Functions and Instruments (Assumption SI-RF):

$$P[y(\cdot)|x = \mathrm{x}, v] = P[y(\cdot)| x = \mathrm{x}]. \tag{7.23}$$

Assumption SI-RF strengthens assumption SI. The latter assumption, when applied to all treatments, asserts that each component $[y(t), t \in T]$ of the outcome vector $y(\cdot)$ is statistically independent of v. Assumption SI-RF asserts that $[y(t), t \in T]$ are jointly independent of v.

The prominence of assumption SI-RF derives from its credibility when the study population are subjects in a randomized experiment. In a randomized experiment, the instrumental variable v designates the treatment group in which each subject has been placed; thus $V = T$. Randomization implies that $y(\cdot)$ is statistically independent of the designated treatment v, so assumption SI-RF holds.

The classical theory of randomized experiments assumes that all subjects comply with their designated treatments; that is, $z = v$. In this special case, application of Proposition 7.3 to each treatment t shows that $P[y(t)|x = \mathrm{x}]$ is point-identified, with

$$H_{SI}\{P[y(t)|x = \mathrm{x}]\} = P(y|x = \mathrm{x}, z = t). \tag{7.24}$$

This finding only uses assumption SI; it does not require the full force of assumption SI-RF.

When some subjects do not comply, randomization of designated treatments generally does not point-identify $P[y(t)|x = \mathrm{x}]$. In this case, assumption SI-RF may have identifying power beyond that obtained when assumption SI is applied to all treatments. However, the form of identification regions under assumption SI-RF is largely an open question.[2]

Complement 7A. Identification and Ambiguity

The treatment-choice problem examined in this chapter is an instance of choice under *ambiguity*. In general, a decision maker with a known choice set who wants to maximize an unknown objective function is said to face a problem of choice under ambiguity. A common source of ambiguity is partial knowledge of a probability distribution describing a relevant population—the decision maker may know only that the distribution of interest is a member of some set of distributions. This is the generic situation of a decision maker who seeks to learn a population distribution empirically, but whose data and other information do not point identify the distribution. Thus, identification problems in empirical analysis generate problems of choice under ambiguity.

The term *ambiguity* appears to originate in Ellsberg (1961), which posed thought experiments in which subjects were asked to draw a ball from either of two urns, one with a known distribution of colors and the other with an unknown distribution of colors. Keynes (1921) and Knight (1921) used the term *uncertainty*, but uncertainty has since come to be used to describe optimization problems in which the objective function depends on a known probability distribution.

Dominated Treatment Rules

Manski (2000) shows that the social planner of Section 7.2 faces a problem of treatment choice under ambiguity whenever identification problems prevent the planner from knowing enough about mean treatment response to be able to determine the optimal rule.

Considering the matter in abstraction, suppose that a planner learns from the available studies that $E[y(\cdot)|x]$ lies in some identification region $H\{E[y(\cdot)|x]\}$. This information may not suffice to solve problem (7.8) but may suffice to determine that some treatment rules are dominated.

A feasible treatment rule $z(\cdot)$ is dominated if there exists another feasible rule, say $z'(\cdot)$, that necessarily yields at least the social welfare of $z(\cdot)$ and that performs strictly better than $z(\cdot)$ in some possible state of nature. Thus $z(\cdot) \in Z(X)$ is dominated if there exists a $z'(\cdot) \in Z(X)$ such that

$$\sum_{x \in X} P(x = x)\{\sum_{t \in T} [\eta(t, x) + c(t, x)] \cdot 1[z(x) = t]\}$$

$$\leq \sum_{x \in X} P(x = x)\{\sum_{t \in T} [\eta(t, x) + c(t, x)] \cdot 1[z'(x) = t]\}$$

for all $\eta \in H\{E[y(\cdot)|x]\}$ and

$$\sum_{x \in X} P(x = x)\{\sum_{t \in T} [\eta(t, x) + c(t, x)] \cdot 1[z(x) = t]\}$$

$$< \sum_{x \in X} P(x = x)\{\sum_{t \in T} [\eta(t, x) + c(t, x)] \cdot 1[z'(x) = t]\}$$

for some $\eta \in H\{E[y(\cdot)|x]\}$, where $\eta(t, x)$ is a feasible value of $E[y(t)|x]$.

Choice Among Undominated Rules

The central difficulty of choice under ambiguity is that there is no clearly best way to choose among undominated actions. Two common suggestions are application of the *maximin rule* or a *Bayes decision rule*.

A planner using the maximin rule selects a treatment rule that maximizes the minimum mean welfare attainable under all possible states of nature. This means solution of the optimization problem

$$\max_{z(\cdot) \in Z^*(X)} \quad \min_{\eta \, \in \, H\{E[y(\cdot)|x]\}} \quad \sum_{x \in X} P(x = x)\{\sum_{t \in T} [\eta(t, x) + c(t, x)] \cdot 1[z(x) = t]\},$$

where $Z^*(X)$ denotes the set of undominated treatment rules.

Bayesian decision theorists recommend that a decision maker facing ambiguity place a subjective distribution on the states of nature and maximize expected welfare with respect to this distribution. In the treatment choice context, the planner would place a probability measure π on the set $H\{E[y(\cdot)|x]\}$, where π expresses the decision maker's personal beliefs about where $E[y(\cdot)|x]$ may lie within $H\{E[y(\cdot)|x]\}$. The planner would then solve the optimization problem

$$\max_{z(\cdot) \, \in \, Z^*(X)} \quad \int \sum_{x \in X} P(x = x)\{\sum_{t \in T} [\eta(t, x) + c(t, x)] \cdot 1[z(x) = t]\} \, d\pi.$$

The maximin rule and Bayes rules are reasonable ways to make decisions under ambiguity, but there is no optimal way to behave in the absence of credible information on the location of $E[y(\cdot)|x]$ within $H\{E[y(\cdot)|x]\}$. Wald (1950), who proposed and studied the maximin rule, did not contend that the rule is optimal, only that it is reasonable. Considering the case in which the objective is to minimize rather than maximize an objective function, he wrote (Wald, 1950, p.18): "a minimax solution seems, in general, to be a reasonable solution of the decision problem."

Bayesians often present *procedural rationality* arguments for use of Bayes decision rules. Savage (1954) showed that a decision maker whose choices are consistent with a certain set of axioms can be interpreted as using a Bayes rule. Many decision theorists consider the Savage axioms, or other sets of axioms, to be a priori appealing. Acting in a manner that is consistent with these axioms does not, however, imply that chosen actions yield good outcomes. Berger (1985, p. 121) calls attention to this, stating: "A Bayesian analysis may be 'rational' in the weak axiomatic sense, yet be terrible in a practical sense if an inappropriate prior distribution is used."

Complement 7B: Sentencing and Recidivism

The question of how judges should sentence convicted juvenile offenders has long been of interest to policy makers and criminologists. Manski and

Nagin (1998) analyzed data on the sentencing of 13,197 juvenile offenders in Utah and their subsequent recidivism. We compared recidivism under the two main sentencing options available to judges: confinement in residential facilities (t = 1) and sentences that do not involve confinement (t = 0).

Let the outcome take the value $y = 1$ if an offender is convicted of a subsequent crime in the two-year period following sentencing, and the value $y = 0$ otherwise. The empirical distribution of treatments and outcomes among the observed offenders was found to be as follows:

residential treatment: $P(z = 1) = 0.11$,
recidivism given residential treatment: $P(y = 1 | z = 1) = 0.77$,
recidivism given nonresidential treatment: $P(y = 1 | z = 0) = 0.59$.

The problem is to use this empirical evidence to draw conclusions about the response probabilities $P[y(1) = 1]$ and $P[y(0) = 1]$.

The empirical evidence alone reveals that

$$P[y(1) = 1] \in [0.08, \ 0.97] \qquad P[y(0) = 1] \in [0.53, 0.64].$$

If one assumes that judges sentence offenders randomly, then

$$P[y(1) = 1] = 0.77 \qquad P[y(0) = 1] = 0.59.$$

Random sentencing did not seem a credible assumption, so we considered two alternative models of treatment selection. The *outcome optimization* model assumes that judges aim to minimize the chance of recidivism. The empirical evidence plus this assumption can be shown to imply that

$$P[y(1) = 1] \in [0.61, \ 0.97] \qquad P[y(0) = 1] \in [0.61, 0.64].$$

The *skimming* model assumes that judges classify offenders as "higher risk" or "lower risk," sentencing only the former to residential confinement. The empirical evidence plus this assumption can be shown to imply that

$$P[y(1) = 1] \in [0.08, \ 0.77] \qquad P[y(0) = 1] \in [0.59, 0.64].$$

Thus, conclusions about response to treatment depend critically on the assumptions imposed.

Complement 7C. Missing Outcome and Covariate Data

Studies of treatment response may have missing data for reasons other than the selection problem. Researchers performing randomized experiments may encounter data collection problems at the beginning of a trial that result in missing covariate or treatment data for some subjects. Subsequently, attrition of subjects may prevent observation of some outcome realizations. Similar problems occur in observational studies, where data on covariates, treatments, or outcomes may be missing due to survey nonresponse.

The analysis of Sections 7.3 and 7.4 assumed that the empirical evidence reveals the distribution $P(y, z, x)$ of (outcomes, treatments, covariates) under the status quo treatment rule. The empirical evidence only partly identifies this distribution when data are missing. Hence, missing data exacerbate the planner's problem.

In principle, it is easy to see how the selection problem and other missing-data problems combine to determine the identification region for $E[y(\cdot)|x]$. Consider the situation using the empirical evidence alone; similar considerations apply when distributional assumptions are imposed. Let $H[P(y, z, x)]$ denote the identification region for $P(y, z, x)$ when some data are missing; as shown in Chapter 3, the particular form of this region depends on the missing data pattern.[3] By (7.11), each feasible distribution $\eta \in H[P(y, z, x)]$ generates an identification region for $E[y(\cdot)|x]$ that recognizes only the selection problem; that is, a region computed under the assumption that $P(y, z, x) = \eta$. Call this region $H_\eta\{E[y(\cdot)|x]\}$. The actual identification region for $E[y(\cdot)|x]$ must recognize the selection problem and other missing-data problems. This region is $\cup_{\eta \in H\{P(y, z, x)\}} H_\eta\{E[y(\cdot)|x]\}$.

In practice, determination of the identification region for $E[y(\cdot)|x]$ when data are missing may be easy or difficult, depending on the specifics of the situation. I provide an empirical illustration in the relatively simple context of a classical randomized experiment, where mean treatment response would be point-identified if all realizations of (y, z, x) were observed.

Choosing Treatments for Hypertension

Physicians routinely face the problem of choosing treatments for hypertension. Medical research has sought to provide guidance through the conduct of randomized trials comparing alternative treatments. Such trials inevitably have missing data. I illustrate here how physicians might use the data from a recent trial to inform treatment choice, without imposing assumptions about the distribution of the missing data.

Materson *et al.* (1993) and Materson, Reda, and Cushman (1995) present findings from a U.S. Department of Veterans Affairs (DVA) trial of treatments for hypertension. Male veteran patients at 15 DVA hospitals were randomly assigned to one of six antihypertensive drug treatments or to a placebo: hydrochlorothiazide (t = 1), atenolol (t = 2), captopril (t = 3), clonidine (t = 4), diltiazem (t = 5), prazosin (t = 6), placebo (t = 7). The trial had two phases. In the first, the dosage that brought diastolic blood pressure (DBP) below 90 mm Hg was determined. In the second, it was determined whether DBP could be kept below 95 mm Hg for a long time. Treatment was defined to be successful if DBP < 90 mm Hg on two consecutive measurement occasions in the first phase and DBP ≤ 95 mm Hg in the second. Treatment was unsuccessful otherwise. Thus, the outcome of interest is binary, with $y = 1$ if the criterion for success is met and $y = 0$ otherwise. Materson et al. (1993) recommend that physicians making treatment choices consider this medical outcome variable as well as a patient's quality of life and the cost of treatment.

Among the covariates measured at the time of randomization, one was the biochemical indicator "renin response," taking the values x = (low, medium, high). This covariate had previously been studied as a factor that might be related to the probability of successful treatment (Freis, Materson, and Flamenbaum, 1983). Renin-response data were missing for some patients in the trial. Moreover, some patients dropped out of the trial before their outcomes could be determined. The pattern of missing covariate and outcome data is shown in Table 1 of Horowitz and Manski (2000), reproduced here.

Table 7C.1 Missing Data in the DVA Hypertension Trial

Treatment	Number Randomized	Observed Successes	None Missing	Missing Only y	Missing Only x	Missing y and x
1	188	100	173	4	11	0
2	178	106	158	11	9	0
3	188	96	169	6	13	0
4	178	110	159	5	13	1
5	185	130	164	6	14	1
6	188	97	164	12	10	2
7	187	57	178	3	6	0

For each value of x, Horowitz and Manski (2000) estimated the identification region for $\{P[y(t) = 1 | x = x], t = 1, \ldots, 7\}$ using the empirical evidence alone. Rather than report the identification regions for these success probabilities, we reported the implied regions for the average treatment effects $\{P[y(t) = 1 | x = x] - \{P[y(7) = 1 | x = x], t = 1, \ldots, 6\}$, which measure the efficacy of each treatment relative to the placebo. This reporting decision was motivated by the traditional research problem of testing the hypothesis of zero average treatment effect. We did not explicitly examine the implications for treatment choice.

Table 7C.2 reports the estimates of the identification regions for the success probabilities themselves. To keep attention focused on the identification problem, suppose that the estimates are the actual identification regions rather than finite sample estimates. Consider a physician who accepts the DVA success criterion, observes renin response, and has no prior information on mean treatment response or the distribution of missing data. Suppose that all treatments have the same cost. How might this physician choose treatments in a population similar to that studied in the DVA trial?

Table 7C.2: Identification Regions for Success Probabilities Conditional on Renin Response

Renin Response	Treatment						
	1	2	3	4	5	6	7
Low	[.54, .61]	[.52, .62]	[.43, .53]	[.58, .66]	[.66, .76]	[.54, .65]	[.29, .32]
Med	[.47, .62]	[.60, .74]	[.53, .68]	[.50, .69]	[.68, .85]	[.41, .65]	[.27, .32]
High	[.28, .50]	[.64, .86]	[.56, .75]	[.63, .84]	[.55, .78]	[.34, .59]	[.28, .40]

The physician should eliminate from consideration the dominated treatments. For patients with low renin response, treatments 1, 2, 3, 4, 6, and 7 are all dominated by treatment 5, which has the greatest lower bound (.66). For patients with medium renin response, treatments 1, 3, 6, and 7 are dominated by treatment 5, which again has the greatest lowest bound (.68). For patients with high renin response, treatments 1, 6, and 7 are dominated by treatment 2, which has the greatest lowest bound in this case (.64). Thus, without imposing any distributional assumptions, the physician can reject treatments 1, 6, and 7 for all patients, can reject treatment 3 for patients with medium renin response, and can determine that treatment 5 is optimal for patients with low renin response.

In the absence of assumptions about the distribution of missing data, it is not possible to give the physician guidance on how to choose among undominated treatments for patients with medium and high renin response. A physician using the maximin rule would choose treatment 5 for patients with medium renin response and treatment 2 for patients with high renin response. This is a reasonable treatment rule, but one cannot say that it is an optimal rule.

Complement 7D. Study and Treatment Populations

A longstanding issue in the analysis of treatment response concerns the importance of correspondence between the study population and the treatment population. This matter was downplayed in the influential work of Donald Campbell, who argued that studies of treatment response should be judged primarily by their internal validity and only secondarily by their external validity (e.g., Campbell and Stanley, 1963; Campbell, 1984). Campbell's view has been endorsed by Rosenbaum (1999), who recommends that observational studies of human subjects aim to approximate the conditions of laboratory experiments (p. 263):

> "In a well-conducted laboratory experiment one of the rarest of things happens: The effects caused by treatments are seen with clarity. Observational studies of the effects of treatments on human populations lack this level of control but the goal is the same. Broad theories are examined in narrow, focused, controlled circumstances."

Rosenbaum, like Campbell, downplays the importance of having the study population be similar to the population of interest, writing (p.259):

> "Studies of samples that are representative of populations may be quite useful in describing those populations, but may be ill-suited to inferences about treatment effects."

From the perspective of treatment choice, the Campbell–Rosenbaum position is well-grounded if treatment response is homogeneous across persons. Then researchers can aim to learn about treatment response in easy-to-analyze study populations and planners can be confident that research findings can be extrapolated to populations of interest. In human populations, however, homogeneity of treatment response may be the exception rather than the rule. Whether the context be medical, educational or social, it is common to find that people vary in their response to treatment. To the degree that treatment response is heterogeneous, a planner cannot readily extrapolate research findings from a study population to a treatment population, as optimal treatments in the two may differ. Hence correspondence between the study population and the treatment population assumes considerable importance.

A specific instance of the general issue arises in research on partial compliance in randomized experiments. Suppose that the study population is formed by drawing experimental subjects at random from the treatment population and randomly designating the treatments they should receive. When some subjects do not comply with their designated treatments,

Imbens and Angrist (1994) and Angrist, Imbens, and Rubin (1996) have proposed that treatment effects be reported for the sub-population of "compliers," persons who would comply with their designated experimental treatments whatever they might be. A planner can extrapolate findings on treatment effects for compliers to the treatment population if treatment response is homogeneous but not to the degree that it is heterogeneous. Indeed, a planner cannot even use findings for compliers to make treatment choices in this particular subpopulation. The reason is that compliers are not individually identifiable. Each subject in an experiment is placed in one of a set of mutually exclusive treatment groups; hence it is not possible to observe whether a given person would comply with all possible treatment designations.

From the perspective of treatment choice in heterogeneous populations, I see no reason to give internal validity primacy relative to external validity. I am unable to motivate interest in the sub-population of compliers. To be fair, researchers who have stressed internal validity and those who have focused attention on compliers have not necessarily asserted that the objective of their research is to inform treatment choice. For example, Angrist, Imbens, and Rubin (1996) view their goal as the discovery of "causal effects," without reference to a treatment-choice problem.

Endnotes

Sources and Historical Notes

This chapter paraphrases and extends ideas introduced in Manski (1990, 2000, 2002).

A large and diverse literature on the evaluation of social programs seeks to compare the outcomes that a population of interest would experience if the members of the population were to receive alternative treatments. The idea of a planner who must choose a treatment rule is implicit in much of this literature, but explicit consideration of the planner's decision problem has been rare until recently. Stafford (1985, pp. 112–114) was an early proponent of the idea.

This chapter views the matter from the perspective of welfare economics, in which a planner aims to maximize a utilitarian social welfare function. Some research on program evaluation adopts a different perspective, in which the aim is to compare a specified "base" or "default" treatment rule z_1 with an alternative z_2. The object of interest is $P[y(z_2) - y(z_1)]$, which measures the distribution of changes in outcomes that would be experienced if the base treatment rule z_1 were to be replaced by the proposed rule z_2. For example, Heckman, Smith, and Clements (1997) write: "Answers to many interesting evaluation questions require knowledge of the distribution of program gains."

Text Notes

1. Although the planner cannot systematically differentiate among persons with the same observed covariates, he can randomly assign different treatments to such persons. Thus, the set of feasible treatment rules in principle contains not only functions mapping covariates into treatments but also probability mixtures of these functions. Explicit consideration of randomized treatment rules would not substantively change the present analysis, but would complicate the necessary notation. A simple implicit way to permit randomized rules is to include in x a component whose value is randomly drawn by the planner from some distribution. The planner can then make the chosen treatment vary with this covariate component.

2. Analysis to date has been limited to the special case in which the outcome is a binary random variable. Let $Y = \{0, 1\}$. Robins (1989) posed the question in this case but only obtained an outer identification region for $\{P[y(0)], P[y(1)]\}$. Balke and Pearl (1997) showed that the identification region under assumption SI-RF is the set of solutions to a certain linear programming problem. They presented numerical examples in which this region sometimes (but not always) is smaller than the one obtained using assumption SI, which is equivalent to assumption MI when response is binary.

3. Chapter 3 covers cases in which outcome and/or covariate data are missing but not ones in which treatment data are missing. Molinari (2002) studies this problem.

8

Monotone Treatment Response

8.1. Shape Restrictions

Empirical researchers studying treatment response sometimes have credible information about the shape of the response functions $y(\cdot)$. In particular, one may have reason to believe that outcomes vary monotonically with the intensity of the treatment. Let the set T of treatments be ordered in terms of degree of intensity. The assumption of *monotone treatment response* asserts that, for all persons j and all treatment pairs (s, t),

$$t \geq s \implies y_j(t) \geq y_j(s). \tag{8.1}$$

This chapter studies the selection problem when response functions are assumed to be monotone in treatments (Section 8.2) or to obey the related shape restrictions of semi-monotonicity (Section 8.3) or concave monotonicity (Section 8.4). Nothing is assumed about the process of treatment selection in the study population, and no cross-person restrictions on response are imposed. The findings reported here may be useful to a planner making treatment choices, as described in Chapter 7, but the present analysis does not presume that the objective is to solve a planning problem. The purpose may simply be to learn about the distribution of treatment response in the study population.

Production Analysis

There are many applied settings in which a planner or researcher may be confident that response is monotone, semi-monotone, or concave-monotone, but be wary of assuming anything else. Economic analysis of production provides a good illustration.

Production analysis typically supposes that firms or other entities use inputs to produce a scalar output; thus, the input is the treatment and the output is the response. Firm j has a *production function* $y_j(\cdot)$ mapping inputs into product output, so $y_j(t)$ is the output that firm j produces when t is the input vector. The most basic tenet of the economic theory of production is that output weakly increases with the level of the inputs. If there is a single input (say labor), this means that treatment response is monotone. If there is a vector of inputs (say labor and capital), treatment response is semi-monotone.

Formally, suppose that there are K inputs, and let $s \equiv (s_1, s_2, \ldots, s_K)$ and $t \equiv (t_1, t_2, \ldots, t_K)$ be two input vectors. Production theory predicts that $y_j(t) \geq y_j(s)$ if input vector t is at least as large, component-by-component, as input vector s; that is, if $t_k \geq s_k$, all $k = 1, \ldots, K$. Production theory does not predict the ordering of $y_j(t)$ and $y_j(s)$ when the input vectors t and s are unordered, each having some components larger than the other. Thus, production functions are semi-monotone.

Consider, for example, the production of corn. The inputs include land and seed. The output is bushels of corn. Production theory predicts that the quantity of corn produced by a farm weakly increases with its input of land and seed. Production theory does not predict the effect on corn production of increasing one input component and decreasing the other.

Economists typically assume more than that production functions are semi-monotone. They often assume that production functions exhibit *diminishing marginal returns*, which means that the production function is concave in each input component, holding the other components fixed. In so-called *short-run* production analysis, researchers distinguish two types of inputs: *variable inputs* whose values can be changed and *fixed inputs* whose values cannot be changed. Short-run production analysis performs thought experiments in which the variable inputs are varied and the fixed inputs are held fixed at their realized values. Thus, the variable inputs are considered to be treatments, the fixed inputs to be covariates, and the short-run production function maps variable inputs into output. Suppose that there is one variable input. Then it is common to assume that the short-run production function $y_j(\cdot)$ is concave-monotone in this input. In the short-run production of corn, for example, seed would usually be thought of as the variable input and land as the fixed input. A researcher might find it plausible to assume that, holding land fixed, output of corn rises with the input of seed but with diminishing returns.

Economists studying production often find it difficult to justify assumptions on the shapes of production functions that go beyond concave monotonicity. Empirical researchers may employ tight parametric models of production, but they rarely can do more than assert on faith that such models are adequate "approximations" to actual production functions. Economists studying production also find it difficult to justify other assumptions that may have identifying power, such as the assumptions using instrumental variables that were discussed in Section 7.4. In particular, it usually makes little economic sense to assume that the input vectors chosen by firms are randomly selected.

D-Outcomes and D-Treatment Effects

The shape restrictions studied in this chapter have particular power to identify parameters of outcome distributions that respect stochastic dominance. Let $D(\cdot)$ be such a parameter and let t be a treatment. Propositions developed below give sharp lower and upper bounds for the *D-outcome* $D[y(t)]$ when treatment response is assumed to be monotone, semi-monotone, or concave-monotone. Corollaries apply the findings to specific D-parameters, including quantiles and means of increasing functions of outcomes. The sharp bounds obtained here are, by definition, the endpoints of the identification regions for the parameters of interest under the maintained shape restrictions. The propositions do not assert that an identification region is the entire interval connecting its endpoints. This is so for expectations but not necessarily for other D-parameters.

Bounds are also obtained for two types of *D-treatment effects*, namely $D[y(t)] - D[y(s)]$ and $D[y(t) - y(s)]$, where $t \in T$ and $s \in T$ are specified treatments. These two treatment effects coincide when $D(\cdot)$ is the expectation functional but otherwise may differ. To distinguish them, $D[y(t)] - D[y(s)]$ is henceforth called a *ΔD-treatment effect* and $D[y(t) - y(s)]$ a *DΔ-treatment effect*. In all cases but one (Proposition 8.9), the reported bounds are sharp.

The propositions developed below apply immediately to inference on conditional D-outcomes and D-treatment effects of the form $D[y(t)|x]$, $D[y(t)|x] - D[y(s)|x]$, and $D[y(t) - y(s)|x]$, where x is an observable covariate. One simply needs to redefine the population of interest to be the sub-population of persons who share a specified value of x. To simplify notation, the analysis does not explicitly condition on x.

Throughout this chapter, the outcome space Y is a closed subset of the extended real line. Whereas the set T of treatments was assumed to be finite in most of Chapter 7, this cardinality assumption is not maintained here. In Section 8.2, T is an ordered set of arbitrary cardinality. In Section 8.3, T is a semi-ordered set of arbitrary cardinality. The treatment set has more structure in Section 8.4, where it is a closed interval on the real line.

8.2. Monotonicity

This section assumes that response is weakly increasing in treatments, as stated in equation (8.1). With obvious modifications, the findings apply when response is weakly decreasing in treatments. The analysis first develops sharp bounds for D-outcomes and then for D-treatment effects.

D-Outcomes

Proposition 8.1 presents the sharp bound on D-outcomes under the assumption of monotone treatment response.

Proposition 8.1: Let T be ordered. Let $y_j(\cdot)$, $j \in J$ be weakly increasing on T. Define

$$y_{0j}(t) \equiv y_j \ \text{if } t \geq z_j$$
$$\phantom{y_{0j}(t)} \equiv y_0 \ \text{otherwise,} \tag{8.2a}$$

$$y_{1j}(t) \equiv y_j \ \text{if } t \leq z_j$$
$$\phantom{y_{1j}(t)} \equiv y_1 \ \text{otherwise.} \tag{8.2b}$$

Then, for every $t \in T$,

$$D[y_0(t)] \leq D[y(t)] \leq D[y_1(t)]. \tag{8.3}$$

This bound is sharp.   □

Proof: Monotonicity of $y_j(\cdot)$ implies this sharp bound on $y_j(t)$:

$$t < z_j \Rightarrow y_0 \leq y_j(t) \leq y_j,$$
$$t = z_j \Rightarrow y_j(t) = y_j, \tag{8.4}$$
$$t > z_j \Rightarrow y_j \leq y_j(t) \leq y_1.$$

Equivalently,

$$y_{0j}(t) \leq y_j(t) \leq y_{1j}(t). \tag{8.5}$$

There are no cross-person restrictions, so the sharp bound on $\{y_j(t), j \in J\}$ is

$$y_{0j}(t) \leq y_j(t) \leq y_{1j}(t), \ j \in J. \tag{8.6}$$

Hence, the random variable $y_0(t)$ is stochastically dominated by $y(t)$, which in turn is stochastically dominated by $y_1(t)$. This shows that (8.3) is a bound on $D[y(t)]$.

The bound (8.3) is sharp because the bound (8.6) is sharp; that is, the empirical evidence and prior information are consistent with the hypothesis $\{y_j(t) = y_{0j}(t), j \in J\}$ and also with the hypothesis $\{y_j(t) = y_{1j}(t), j \in J\}$.

Q. E. D.

Proposition 8.1 shows that the assumption of monotone treatment response qualitatively reduces the severity of the selection problem. Using the empirical evidence alone, an observed outcome realization y_j is informative about outcome $y_j(t)$ only if $z_j = t$; then $y_j = y_j(t)$. Using the empirical evidence and the monotone-response assumption, observation of y_j always yields an informative lower or upper bound on $y_j(t)$, as shown in equation (8.4).

Proposition 8.1 is simple to state and prove, but it is too abstract to give a clear sense of the identifying power of the monotonicity assumption. This emerges in Corollaries 8.1.1 through 8.1.3, which apply the proposition to upper tail probabilities, the means of increasing functions, and quantiles. In each case, the corollary is obtained by evaluating $D[y_0(t)]$ and $D[y_1(t)]$ for the D-parameter of interest.

Corollary 8.1.1: Let $f(\cdot): R \rightarrow R$ be weakly increasing. Then

$$f(y_0)P(t < z) + E[f(y) \mid t \geq z] \cdot P(t \geq z) \leq E\{f[y(t)]\}$$

$$\leq f(y_1)P(t > z) + E[f(y) \mid t \leq z] \cdot P(t \leq z). \tag{8.7}$$

□

Corollary 8.1.2: Let $\alpha \in (0, 1)$. Let $Q_\alpha(u)$ denote the α–quantile of a real random variable u. Let $\lambda_0 \equiv [\alpha - P(t < z)]/P(t \geq z)$ and $\lambda_1 \equiv \alpha/P(t \leq z)$. Then

$$0 < \alpha \leq P(t < z) \Rightarrow y_0 \leq Q_\alpha[y(t)] \leq Q_{\lambda 1}(y \mid t \leq z),$$

$$P(t < z) < \alpha \leq P(t \leq z) \Rightarrow Q_{\lambda 0}(y \mid t \geq z) \leq Q_\alpha[y(t)] \leq Q_{\lambda 1}(y \mid t \leq z), \tag{8.8}$$

$$P(t \leq z) < \alpha < 1 \Rightarrow Q_{\lambda 0}(y \mid t \geq z) \leq Q_\alpha[y(t)] \leq y_1.$$

□

It is revealing to compare Corollary 8.1.1, which exploits the assumption of monotone response, with the bound on $E\{f[y(t)]\}$ using the empirical evidence alone. The bound using the empirical evidence alone is

$$f(y_0)P(t \neq z) + E[f(y) \mid t = z] \cdot P(t = z) \leq E\{f[y(t)]\}$$

$$\leq f(y_1)P(t \neq z) + E[f(y) \mid t = z] \cdot P(t = z). \tag{8.9}$$

Whereas this bound draws information about $E\{f[y(t)]\}$ only from the (y, z) pairs with $t = z$, all (y, z) pairs are informative under the monotone-response assumption. The lower bound in Corollary 8.1.1 draws information from the persons with $t \geq z$, and the upper bound draws information from those with $t \leq z$.

Corollary 8.1.1 may be used to obtain sharp bounds on tail probabilities. Let $r \in (y_0, y_1]$. The indicator function $1[y(t) \geq r]$ is an increasing function of $y(t)$ and $E\{1[y(t) \geq r]\} = P[y(t) \geq r]$, so inequality (8.7) reduces to

$$P(t \geq z \cap y \geq r) \leq P[y(t) \geq r] \leq P(t > z \cup y \geq r). \tag{8.10}$$

The informativeness of this bound depends on the distribution $P(y, z)$ of realized treatments and outcomes. Suppose that $P(t \geq z \cap y \geq r) = 0$ and $P(t > z \cup y \geq r) = 1$. Then (8.10) is the trivial bound $0 \leq P[y(t) \geq r] \leq 1$. Suppose however that $P(t \geq z \cap y \geq r) = P(t > z \cup y \geq r)$. Then the assumption of monotone treatment response point-identifies $P[y(t) \geq r]$.

Corollary 8.1.2 shows that the monotone response generically yields a one-sided bound on quantiles of $y(t)$. The upper bound is informative when $\alpha \leq P(t \leq z)$. The lower bound is informative when $\alpha > P(t < z)$. These cases exhaust the possibilities if $P(t = z) = 0$. The lower and upper bounds are both informative if $P(t = z) > 0$ and $P(t < z) < \alpha \leq P(t \leq z)$.

D-Treatment Effects

Propositions 8.2 and 8.3 present the sharp bounds on the D-treatment effects $D[y(t)] - D[y(s)]$ and $D[y(t) - y(s)]$.

Proposition 8.2: Let T be ordered. Let $y_j(\cdot)$, $j \in J$ be weakly increasing on T. Then for every $t \in T$ and $s \in T$ with $t > s$,

$$0 \leq D[y(t)] - D[y(s)] \leq D[y_1(t)] - D[y_0(s)]. \tag{8.11}$$

This bound is sharp.   □

Proof: Monotonicity of response implies that $y(t)$ stochastically dominates $y(s)$, so 0 is a lower bound on $D[y(t)] \quad D[y(s)]$. Proposition 8.1 implies that

$D[y_1(t)] - D[y_0(s)]$ is an upper bound. We need to prove that these bounds are sharp.

Let $j \in J$. Monotonicity of $y_j(\cdot)$ gives this sharp bound on $\{y_j(t), y_j(s)\}$:

$$s < t < z_j \;\Rightarrow\; y_0 \le y_j(s) \le y_j(t) \le y_j,$$

$$s < t = z_j \;\Rightarrow\; y_0 \le y_j(s) \le y_j(t) = y_j,$$

$$s < z_j < t \;\Rightarrow\; y_0 \le y_j(s) \le y_j \le y_j(t) \le y_1, \qquad (8.12)$$

$$s = z_j < t \;\Rightarrow\; y_j = y_j(s) \le y_j(t) \le y_1,$$

$$z_j < s < t \;\Rightarrow\; y_j \le y_j(s) \le y_j(t) \le y_1.$$

There are no cross-person restrictions, so the empirical evidence and prior information are consistent with the hypothesis $\{y_j(s) = y_j(t), j \in J\}$ and also with the hypothesis $\{y_j(t) = y_{1j}(t), y_j(s) = y_{0j}(s), j \in J\}$. Hence (8.11) is sharp.

Q. E. D.

Proposition 8.3: Let T be ordered. Let $y_j(\cdot)$, $j \in J$ be weakly increasing on T. Then for every $t \in T$ and $s \in T$ with $t > s$,

$$D(0) \;\le\; D[y(t) - y(s)] \;\le\; D[y_1(t) - y_0(s)]. \qquad (8.13)$$

This bound is sharp.   □

Proof: The proof to Proposition 8.2 showed that the sharp joint bound on $\{y_j(t) - y_j(s), j \in J\}$ is

$$0 \;\le\; y_j(t) - y_j(s) \;\le\; y_{1j}(t) - y_{0j}(s), \quad j \in J. \qquad (8.14)$$

Hence, the degenerate distribution with all mass at 0 is stochastically dominated by $y(t) - y(s)$, which in turn is dominated by $y_1(t) - y_0(s)$. Thus, (8.13) is a bound on $D[y(t) - y(s)]$. This bound is sharp because bound (8.14) is sharp.

Q. E. D.

Observe that the lower bounds in Propositions 8.2 and 8.3, namely 0 and $D(0)$, are implied by the monotone-response assumption and do not depend on the empirical evidence. The monotone-response assumption and the empirical evidence together determine the upper bounds.

Propositions 8.2 and 8.3 generically give distinct bounds for distinct treatment effects, but these bounds coincide when $D(\cdot)$ is the expectation functional. Proposition 8.2 and Corollary 8.1.1 yield the following.

Corollaries 8.2.1 and 8.3.1:

$$0 \leq E[y(t)] - E[y(s)] = E[y(t) - y(s)]$$

$$\leq y_1 \cdot P(t > z) + E(y \mid t \leq z) \cdot P(t \leq z) - y_0 \cdot P(s < z) - E(y \mid s \geq z) \cdot P(s \geq z). \tag{8.15}$$

This bound is sharp.   □

This result takes a particularly simple form when outcomes are binary. Let Y be the two-element set $\{0, 1\}$. Then $y_0 = 0$, $y_1 = 1$, and (8.15) becomes

$$0 \leq P[y(t) = 1] - P[y(s) = 1] = P[y(t) - y(s) = 1]$$

$$\leq P(y = 0, t > z) + P(y = 1, s < z). \tag{8.16}$$

8.3. Semi-Monotonicity

In this section, treatments are K-dimensional vectors and T is a semi-ordered set of treatment vectors. The notation $s \oslash t$ indicates that a pair (s, t) is not ordered. Analysis of semi-monotone response uses much of the structure developed in Section 8.2, and so is amenable to succinct presentation. When T is semi-ordered, the definitions of $y_{0j}(t)$ and $y_{1j}(t)$ in (8.2) remain valid, the term "otherwise" now including the possibility that $t \oslash z_j$.

D-Outcomes

Proposition 8.4 gives the semi-monotone-response version of Proposition 8.1. Observe that the conclusion to Proposition 8.1 still holds and the proof requires only slight modification.

Proposition 8.4: Let T be semi-ordered. Let $y_j(\cdot), j \in J$ be weakly increasing on the ordered pairs in T. Then, for every $t \in T$,

$$D[y_0(t)] \leq D[y(t)] \leq D[y_1(t)]. \tag{8.17}$$

This bound is sharp.   □

Proof: Let $j \in J$. Semi-monotonicity of $y_j(\cdot)$ implies this sharp bound on $y_j(t)$:

$$t < z_j \;\Rightarrow\; y_0 \le y_j(t) \le y_j,$$

$$t = z_j \;\Rightarrow\; y_j(t) = y_j,$$

$$t > z_j \;\Rightarrow\; y_j \le y_j(t) \le y_1, \tag{8.18}$$

$$t \oslash z_j \;\Rightarrow\; y_0 \le y_j(t) \le y_1.$$

Thus (8.5) holds. The rest of the proof is the same as the proof to Proposition 8.1.

Q. E. D.

Although Propositions 8.1 and 8.4 have the same stated conclusion, weakening the assumption of monotone response to semi-monotone response is consequential. Suppose that an ordering of the treatment set T is weakened to a semi-ordering. Each time that a pair (t, z_j) with $t > z_j$ becomes unordered, $y_{0j}(t)$ falls from y_j to y_0. Each time that a pair (t, z_j) with $t < z_j$ becomes unordered, $y_{1j}(t)$ rises from y_j to y_1. Hence, the ordered-T version of $y_0(t)$ stochastically dominates its semi-ordered-T counterpart, and the ordered-T version of $y_1(t)$ is stochastically dominated by its semi-ordered-T counterpart. Thus, weakening the assumption of monotone response to semi-monotone response widens the bound on $D[y(t)]$. In the extreme case where T is entirely unordered, Proposition 8.4 gives the bound on $D[y(t)]$ obtained using the empirical evidence alone.

Corollaries 8.4.1 and 8.4.2 give the semi-monotone-response versions of Corollaries 8.1.1 and 8.1.2. The explicit forms for $D[y_0(t)]$ and $D[y_1(t)]$ in these corollaries show clearly how weakening the assumption of monotone response to semi-monotone response affects the bounds on D-outcomes.

Corollary 8.4.1: Let $f(\cdot): R \to R$ be weakly increasing. Then

$$f(y_0)P(t < z \cup t \oslash z) + E[f(y) \,|\, t \ge z] \cdot P(t \ge z) \;\le\; E\{f[y(t)]\}$$

$$\le\; f(y_1)P(t > z \cup t \oslash z) + E[f(y) \,|\, t \le z] \cdot P(t \le z). \tag{8.19}$$

□

Corollary 8.4.2: Let $\alpha \in (0, 1)$. Let $\lambda_0 \equiv [\alpha - P(t < z \cup t \oslash z)]/P(t \ge z)$ and $\lambda_1 \equiv \alpha/P(t \le z)$. Then

$$0 < \alpha \le P(t < z \cup t \oslash z) \Rightarrow y_0 \le Q_\alpha[y(t)],$$

$$P(t < z \cup t \oslash z) < \alpha < 1 \Rightarrow Q_{\lambda 0}(y \,|\, t \ge z) \le Q_\alpha[y(t)],$$

$$0 < \alpha \le P(t \le z) \Rightarrow Q_\alpha[y(t)] \le Q_{\lambda 1}(y \,|\, t \le z), \qquad (8.20)$$

$$P(t \le z) < \alpha < 1 \Rightarrow Q_\alpha[y(t)] \le y_1.$$

□

Corollary 8.4.1 may be used to obtain sharp bounds on tail probabilities. The result is

$$P(t \ge z \cap y \ge r) \le P[y(t) \ge r] \le P(t > z \cup t \oslash z \cup y \ge r). \qquad (8.21)$$

D-Treatment Effects

When T is semi-ordered, the conclusions to Propositions 8.2 and 8.3 still hold if $t > s$. The upper bounds still hold if $t \oslash s$, but the lower bounds need to be modified. Propositions 8.5 and 8.6 give these extensions to the earlier results.

Proposition 8.5: Let T be semi-ordered. Let $t \in T$ and $s \in T$. Let $y_j(\cdot)$, $j \in J$ be weakly increasing on the ordered pairs in T. For $t > s$, the sharp bound on $D[y(t)] - D[y(s)]$ is

$$0 \le D[y(t)] - D[y(s)] \le D[y_1(t)] - D[y_0(s)]. \qquad (8.22)$$

For $t \oslash s$, the sharp bound is

$$D[y_0(t)] - D[y_1(s)] \le D[y(t)] - D[y(s)] \le D[y_1(t)] - D[y_0(s)]. \qquad (8.23)$$

□

Proof: Let $t > s$. Semi-monotonicity of response implies that $y(t)$ stochastically dominates $y(s)$, so 0 is a lower bound on $D[y(t)] - D[y(s)]$. Proposition 8.4 implies that $D[y_1(t)] - D[y_0(s)]$ is an upper bound. To prove that these bounds are sharp, consider $j \in J$. If s, t, and z_j are ordered, (8.12) still gives the sharp joint bound on $y_j(t)$ and $y_j(s)$. If $z_j \oslash t$ and/or $z_j \oslash s$, the sharp joint bound on $y_j(t)$ and $y_j(s)$ is

$$s < t \cap s < z_j \;\Rightarrow\; y_0 \le y_j(s) \le y_j \cap y_j(s) \le y_j(t) \le y_1,$$

$$s < t \cap z_j < t \;\Rightarrow\; y_0 \le y_j(s) \le y_j(t) \cap y_j \le y_j(t) \le y_1, \qquad (8.24)$$

$$s < t \;\Rightarrow\; y_0 \le y_j(s) \le y_j(t) \le y_1.$$

The rest of the proof is the same as the proof to Proposition 8.2.

Let $s \oslash t$. Proposition 8.4 implies that (8.23) is a bound on $D[y(t)]$ - $D[y(s)]$. For each $j \in J$, the sharp joint bound on $y_j(t)$ and $y_j(s)$ is

$$y_{0j}(t) \le y_j(t) \le y_{1j}(t),$$

$$y_{0j}(s) \le y_j(s) \le y_{1j}(s). \qquad (8.25)$$

There are no cross-person restrictions, so the empirical evidence and prior information are consistent with the hypothesis $\{y_j(t) = y_{0j}(t)$ and $y_j(s) = y_{1j}(s), j \in J\}$ and with the hypothesis $\{y_j(t) = y_{1j}(t)$ and $y_j(s) = y_{0j}(s), j \in J\}$. Hence (8.23) is sharp.

Q. E. D.

Proposition 8.6: Let T be semi-ordered. Let $t \in T$ and $s \in T$. Let $y_j(\cdot), j \in J$ be weakly increasing on the ordered pairs in T. For $t > s$, the sharp bound on $D[y(t) - y(s)]$ is

$$D(0) \le D[y(t) - y(s)] \le D[y_1(t) - y_0(s)]. \qquad (8.26)$$

For $s \oslash t$, the sharp bound is

$$D[y_0(t) - y_1(s)] \le D[y(t) - y(s)] \le D[y_1(t) - y_0(s)]. \qquad (8.27)$$

□

Proof: Let $t > s$. By (8.12) and (8.24), the sharp bound on $\{y_j(t) - y_j(s), j \in J\}$ is

$$0 \le y_j(t) - y_j(s) \le y_{1j}(t) - y_{0j}(s), \; j \in J. \qquad (8.28)$$

The rest of the proof is the same as the proof to Proposition 8.3.

Let $s \oslash t$. By (8.25), the sharp joint bound on $\{y_j(t) - y_j(s), j \in J\}$ is

$$y_{0j}(t) - y_{1j}(s) \le y_j(t) - y_j(s) \le y_{1j}(t) - y_{0j}(s), \; j \in J. \qquad (8.29)$$

Hence (8.27) is the sharp bound on $D[y(t) - y(s)]$.

Q. E. D.

Testing the Hypothesis of Semi-monotone Response

Whereas Propositions 8.4 through 8.6 take semi-monotone treatment response as a maintained assumption, one may instead want to view it as a hypothesis to be tested. It is easy to see that the hypothesis is not refutable in isolation. For each $j \in J$, only one point on the response function $y_j(\cdot)$ is observable, namely $y_j(z_j)$. Hence, the empirical evidence is necessarily consistent with the hypothesis that $y_j(\cdot)$ is weakly increasing on the ordered pairs in T. In particular, the empirical evidence is consistent with the hypothesis that every response function is flat, with $\{y_j(t) = y_j, t \in T, j \in J\}$.

A researcher wanting to test the hypothesis of semi-monotone response can do so only if this hypothesis is joined with other assumptions. Consider assumption SI-RF, which states that z is statistically independent of $y(\cdot)$; that is,

$$P[y(\cdot)] = P[y(\cdot)|z]. \tag{8.30}$$

The joint hypothesis of semi-monotone response and assumption SI-RF is refutable. The key is Proposition 8.7.

Proposition 8.7: Let T be semi-ordered. Let $t > s$. Let $y_j(\cdot), j \in J$ be weakly increasing on the ordered pairs in T. Let z be statistically independent of $y(\cdot)$. Then $P(y|z = t)$ stochastically dominates $P(y|z = s)$. □

Proof: Semi-monotonicity implies that $y(t)$ stochastically dominates $y(s)$. Assumption SI-RF implies that $P[y(s)] = P(y|z=s)$ and $P[y(t)] = P(y|z = t)$.

Q. E. D.

Empirical knowledge of the distribution $P(y, z)$ of realized treatments and outcomes implies knowledge of $P(y|z = s)$ and $P(y|z = t)$ for s and t on the support of $P(z)$, so Proposition 8.7 yields this test:

Reject the joint hypothesis of semi-monotone treatment response and assumption SI-RF if there exist $s \in T$ and $t \in T$ on the support of $P(z)$ such that $t > s$ but $P(y|z = t)$ does not stochastically dominate $P(y|z = s)$.

In finite-sample practice, a researcher observing a random sample of (y, z) pairs can estimate $P(y|z = t)$ and $P(y|z = s)$ and form an asymptotically valid version of the test.

There are three ways to interpret an empirical finding that $P(y|z = t)$ does not stochastically dominate $P(y|z = s)$. Researchers who have confidence in assumption SI-RF would conclude that response is not semi-monotone. Researchers confident that response is semi-monotone would conclude that assumption SI-RF does not hold. Other researchers would conclude only that some part of the joint hypothesis is incorrect.

8.4. Concave Monotonicity

Whereas Section 8.3 weakened the assumption of monotone response to semi-monotonicity, this section strengthens the assumption. In particular, $y_j(\cdot)$, $j \in J$ now are concave-monotone functions. Moreover, $T = [0, \tau]$ for some $\tau \in (0, \infty]$ and $Y = [0, \infty]$. The important feature of this specification of T and Y is that these sets are closed intervals with finite lower bounds. Specifying that the lower bounds of T and Y are zero and that the upper bound of Y is ∞ merely permits some simplification in the analysis.[1]

The analysis uses this fact: Given three points $(v_m, w_m) \in [0, \infty]^2$, $m = 1$, 2, 3, with $0 < v_1 < v_2 < v_3$, there exists a concave-monotone function mapping $[0, \tau] \rightarrow [0, \infty]$ and passing through the three points if and only if

$$w_1/v_1 \geq (w_2 - w_1)/(v_2 - v_1) \geq (w_3 - w_2)/(v_3 - v_2) \geq 0. \qquad (8.31)$$

Here w_1/v_1 is the slope of the line segment connecting the origin to (v_1, w_1), and $(w_m - w_{m-1})/(v_m - v_{m-1})$ is the slope of the line segment connecting (v_{m-1}, w_{m-1}) to (v_m, w_m), $m = 2, 3$. In particular, the piecewise linear function passing through the origin and the three points is concave-monotone if and only if (8.31) holds.

D-Outcomes

Proposition 8.8 presents the sharp bound on D-outcomes under the assumption of concave-monotone treatment response.

Proposition 8.8: Let $T = [0, \tau]$ and $Y = [0, \infty]$. Let $y_j(\cdot)$, $j \in J$ be concave and weakly increasing on T. Define

$$\begin{aligned} y_{c0j}(t) &\equiv y_j \text{ if } t \geq z_j \\ &\equiv y_j t/z_j \text{ otherwise,} \end{aligned} \qquad (8.32a)$$

$$\begin{aligned} y_{c1j}(t) &\equiv y_j \text{ if } t \leq z_j \\ &\equiv y_j t/z_j \text{ otherwise.} \end{aligned} \qquad (8.32b)$$

Then, for every $t \in T$,

$$D[y_{c0}(t)] \leq D[y(t)] \leq D[y_{c1}(t)]. \tag{8.33}$$

This bound is sharp.   □

Proof: For $j \in J$, $y_j(\cdot)$ is a concave-monotone function passing through (z_j, y_j) and $[t, y_j(t)]$. Application of (8.31) yields this sharp bound on $y_j(t)$:

$$t < z_j \Rightarrow y(t)/t \geq [y_j - y(t)]/(z_j - t) \geq 0$$
$$\Rightarrow y_j t/z_j \leq y_j(t) \leq y_j,$$

$$t = z_j \Rightarrow y_j(t) = y_j, \tag{8.34}$$

$$t > z_j \Rightarrow y_j/z_j \geq [y_j(t) - y_j]/(t - z_j) \geq 0$$
$$\Rightarrow y_j \leq y_j(t) \leq y_j t/z_j.$$

Equivalently,

$$y_{c0j}(t) \leq y_j(t) \leq y_{c1j}(t). \tag{8.35}$$

The rest of the proof is the same as the proof to Proposition 8.1, with $y_{c0j}(t)$ and $y_{c1j}(t)$ replacing $y_{0j}(t)$ and $y_{1j}(t)$.

Q. E. D.

Comparison of the bounds $[y_{c0j}(t), y_{c1j}(t)]$ and $[y_{0j}(t), y_{1j}(t)]$ shows that strengthening the assumption of monotone response to concave-monotone response has considerable identifying power. Monotonicity implies that observation of the realized outcome y_j yields either an informative lower or upper bound on $y_j(t)$, as shown in equation (8.4). Concave monotonicity implies that observation of y_j yields both an informative lower and upper bound on $y_j(t)$, as shown in equation (8.34). The present bound on $y_j(t)$ is not only narrower than the earlier one, but its width varies with t in a qualitatively different manner. The present bound has width $y_j \cdot |(z_j - t)/z_j|$, whereas the earlier one has width $y_j \cdot 1[t < z_j] + \infty \cdot 1[t > z_j]$. Thus, the present bound widens linearly from zero as t moves away from z_j, whereas the width of the earlier one varies discontinuously with t.

Corollaries 8.8.1 and 8.8.2 give the concave-monotone response versions of Corollaries 8.1.1 and 8.1.2. Comparison of these corollaries with the earlier ones shows clearly the additional identifying power of assuming that response is concave. The earlier bounds on $E\{f[y(t)]\}$ are uninformative if $f(y_0) = -\infty$ and $f(y_1) = \infty$, but the present bounds are essentially always informative. The earlier bounds on quantiles of $y(t)$ are generically informative only from one side or the other, but the present bounds are informative both from above and below.

Corollary 8.8.1: Let $f(\cdot)$: $R \rightarrow R$ be weakly increasing. Then

$$E[f(yt/z) \mid t < z] \cdot P(t < z) + E[f(y) \mid t \geq z] \cdot P(t \geq z) \leq E\{f[y(t)]\}$$

$$\leq E[f(yt/z) \mid t > z] \cdot P(t > z) + E[f(y) \mid t \leq z] \cdot P(t \leq z). \qquad (8.36)$$

$\square$

Corollary 8.8.2: Let $\alpha \in (0, 1)$. Then

$$Q_\alpha\{y \cdot 1[t \geq z] + yt/z \cdot 1[t < z]\} \leq Q_\alpha[y(t)]$$

$$\leq Q_\alpha\{y \cdot 1[t \leq z] + yt/z \cdot 1[t > z]\}. \qquad (8.37)$$

$\square$

Corollary 8.8.1 implies sharp bounds on tail probabilities. The result is

$$P[(t \geq z \cap y \geq r) \cup (t < z \cap yt/z \geq r)] \leq P[y(t) \geq r]$$

$$\leq P[(t > z \cap yt/z \geq r) \cup (t \leq z \cap y \geq r)]. \qquad (8.38)$$

D-Treatment Effects

Bounds on D-treatment effects follow from the sharp bounds obtained for $\{y_j(t), y_j(s)\}$. In Sections 8.2 and 8.3, we found these joint bounds to have simple forms when response is monotone or semi-monotone. The joint bounds assuming concave-monotone response are more complex. Application of (8.31) yields these bounds on $\{y_j(t), y_j(s)\}$:

$$s < t < z_j \Rightarrow y_j(s)/s \geq [y_j(t) - y_j(s)]/(t - s) \geq [y_j - y_j(t)]/(z_j - t) \geq 0,$$

$$s < t = z_j \Rightarrow y_j(s)/s \geq [y_j(t) - y_j(s)]/(t - s) = [y_j - y_j(s)]/(z_j - s) \geq 0,$$

$$s < z_j < t \Rightarrow y_j(s)/s \geq [y_j - y_j(s)]/(z_j - s) \geq [y_j(t) - y_j]/(t - z_j) \geq 0, \qquad (8.39)$$

$$s = z_j < t \Rightarrow y_j(s)/s = y_j/z_j \geq [y_j(t) - y_j]/(t - z_j) \geq 0,$$

$$z_j < s < t \Rightarrow y_j/z_j \geq [y_j(s) - y_j]/(s - z_j) \geq [y_j(t) - y_j(s)]/(t - s) \geq 0.$$

Proposition 8.10 uses (8.39) to derive the sharp bound on $D[y(t) - y(s)]$. Proposition 8.9 gives the sharp lower bound on $D[y(t)] - D[y(s)]$ but only a non-sharp upper bound.

Proposition 8.9: Let $T = [0, \tau]$ and $Y = [0, \infty]$. Let $y_j(\cdot), j \in J$ be concave and weakly increasing on T. Then, for every $t \in T$ and $s \in T$ with $t > s$,

$$0 \leq D[y(t)] - D[y(s)] \leq D[y_{c1}(t)] - D[y_{c0}(s)]. \qquad (8.40)$$

The lower bound is sharp, but the upper bound is not sharp.   □

Proof: Monotonicity of response implies that $y(t)$ stochastically dominates $y(s)$, so 0 is a lower bound on $D[y(t)] - D[y(s)]$. This lower bound is sharp because the hypothesis $\{y_j(t) = y_j(s) = y_j, j \in J\}$ satisfies (8.39).

Proposition 8.8 implies that $D[y_{c1}(t)] - D[y_{c0}(s)]$ is an upper bound on $D[y(t)] - D[y(s)]$. This upper bound is not sharp because the hypothesis $\{y_j(t) = y_{c1j}(t), y_j(s) = y_{c0j}(s), j \in J\}$ does not satisfy (8.39). When $s < t < z_j$, setting $\{y_j(t) = y_j, y_j(s) = y_j s/z_j\}$ violates (8.39). Similarly, when $z_j < s < t$, setting $\{y_j(t) = y_j t/z_j, y_j(s) = y_j\}$ violates (8.39).

                                                                        Q. E. D.

Proposition 8.10: Let $T = [0, \tau]$. Let $y_j(\cdot), j \in J$ be concave and weakly increasing on T. Let $Y = [0, \infty]$. For each $t \in T$ and $s \in T$ with $t > s$, define

$$
\begin{aligned}
y_{ctj}(s) &\equiv y_j s/t \quad \text{if } t < z_j, \\
&\equiv y_j s/z_j \text{ otherwise.}
\end{aligned}
\qquad (8.41)
$$

Then

$$D(0) \leq D[y(t) - y(s)] \leq D[y_{c1}(t) - y_{ct}(s)]. \qquad (8.42)$$

This bound is sharp.   □

Proof: The lower bound holds because response is monotone. It is sharp because the hypothesis $\{y_j(t) - y_j(s) = 0, j \in J\}$ satisfies (8.39).

To obtain the sharp upper bound, we need to determine the largest value of $y_j(t) - y_j(s)$ that satisfies (8.39). This can be accomplished in two steps. First hold $y_j(t)$ fixed and minimize $y_j(s)$ subject to (8.39). This yields the maximum of $y_j(t) - y_j(s)$ as a function of $y_j(t)$. Then maximize this expression over $y_j(t) \in [y_{c0j}(t), y_{c1j}(t)]$.

When $t < z_j$, setting $\{y_j(t) = y_j, y_j(s) = y_j s/t\}$ yields the maximal value of $y_j(t) - y_j(s)$. When $t \geq z_j$, setting $\{y_j(t) = y_j t/z_j, y_j(s) = y_j s/z_j\}$ yields the maximal value of $y_j(t) - y_j(s)$. It follows that the sharp bound on $\{y_j(t) - y_j(s), j \in J\}$ is

$$0 \leq y_j(t) - y_j(s) \leq y_{c1j}(t) - y_{ctj}(s), \quad j \in J. \tag{8.43}$$

Hence $D[y_{c1}(t) - y_{ct}(s)]$ is the sharp upper bound on $D[y(t) - y(s)]$.

Q. E. D.

Proposition 8.10 may be applied to give the sharp bound for average treatment effects. Write $y_{c1}(t) - y_{ct}(s)$ in the explicit form

$$y_{c1}(t) - y_{ct}(s) = 1[t < z] \cdot (t - s) \cdot y/t + 1[t \geq z] \cdot (t - s) \cdot y/z. \tag{8.44}$$

The result is the following corollary.

Corollary 8.10.1:

$$0 \leq E[y(t)] - E[y(s)] = E[y(t) - y(s)]$$

$$\leq (t - s) \cdot [E(y/t \mid t < z) \cdot P(t < z) + E(y/z \mid t \geq z) \cdot P(t \geq z)]. \tag{8.45}$$

This bound is sharp.   □

Complement 8A: Downward-Sloping Demand

Section 8.1 used the analysis of production to illustrate the shape restrictions studied in this chapter. Another economic illustration is the assumption that demand functions slope downward.

Economic analyses of market demand usually suppose that there is a set of isolated markets for a given product. Each market is characterized by a demand function, which gives the quantity of product that price-taking consumers would purchase if the price were set at any specified level. In each market, the interaction of consumers and firms determines the price at which transactions actually take place.

In the language of the present chapter, markets are persons, prices are treatments, and quantity demanded is an outcome. Thus T is the set of logically possible prices. In each market j, transactions take place at some realized price $z_j \in T$. The market demand function is $y_j(\cdot)$, and $y_j \equiv y_j(z_j)$ is the quantity actually transacted in market j. The empirical evidence is data on the quantities, prices, and covariates (y_i, z_i, x_i), $i = 1, \ldots, N$ realized in a random sample of N markets. The inferential problem is to combine this evidence with prior information to learn about the distribution $P[y(\cdot)]$ of demand functions across markets.

The one relatively firm conclusion of the theory of demand is that market demand ordinarily is a downward-sloping function of price. This is not a universal prediction. Introductory textbooks expositing consumer theory distinguish between the substitution and income effects of price changes. When income effects are sufficiently strong, consumer optimization implies the existence of *Giffen goods*, for which demand increases with price over some domain. The modern theory of markets with imperfect information emphasizes that price may convey information. If the informational content of price is sufficiently strong, demand functions need not always slope downward. These exceptions notwithstanding, the ordinary presumption of economists is that demand functions are downward-sloping.

Economic theory does not yield other conclusions about the shape of demand functions. Nor does the theory of demand imply anything about price determination. Conclusions about price determination can be drawn only if assumptions about the structure of demand are combined with assumptions about the behavior of the firms that produce the product in question. Thus, demand analysis offers a good example of an inferential problem in which the analyst can reasonably assert that response functions are monotone but should be wary of imposing other assumptions.

Oddly, the classical econometric analysis of demand and supply as linear simultaneous equations does not assume that market demand is downward-sloping. Instead, it imposes another assumption on the structure of demand functions. Begun in the 1920s, brought to maturity in Hood and Koopmans (1953), and exposited regularly in subsequent econometrics texts, the classical analysis assumes that demand is a linear function of price, with the same slope parameter in each market. Thus

$$y_j(t) \;=\; \text{ß}t + u_j, \tag{8A.1}$$

where ß is the common slope parameter and u_j is a market-specific intercept. Nothing is assumed about the sign or magnitude of ß.

Economic theory does not suggest that demand should be linear in price, and applied researchers rarely motivate the assumption. The main appeal of (8A.1) is that it reduces the problem of inference on the distribution $P[y(\cdot)]$ of demand functions to one of inference on the scalar parameter ß. The central classical finding is that $P[y(\cdot)]$ is point-identified if (8A.1) is combined with the mean-independence assumption[2]

$$E(u \,|\, v = v_0) \;=\; E(u \,|\, v = v_1), \tag{8A.2a}$$
$$E(z \,|\, v = v_0) \;\neq\; E(z \,|\, v = v_1), \tag{8A.2b}$$

where v is an instrumental variable taking the values v_0 and v_1. Here is a simple proof taken from Manski (1995, p. 152).

Assumption (8A.1) implies that $u_j = y_j - ßz_j$ in each market j. This and (8A.2a) imply that

$$E(y - ßz \mid v = v_0) = E(y - ßz \mid v = v_1).\qquad (8A.3)$$

Solving (8A.3) for ß yields

$$ß = \frac{E(y \mid v = v_0) - E(y \mid v = v_1)}{E(z \mid v = v_0) - E(z \mid v = v_1)}, \qquad (8A.4)$$

provided that (8A.2b) holds. Empirical knowledge of $P(y, z, v)$ identifies the conditional expectations $E(y \mid v)$ and $E(z \mid v)$ on the right side of (8A.4), so ß is point-identified. Knowledge of ß and $P(y, z)$ implies knowledge of $P(u)$ and hence $P[y(\cdot)]$.

Complement 8B. Econometric Response Models

This chapter has assumed only that the response functions $y_j(\cdot), j \in J$ share the common property of monotonicity, semi-monotonicity, or concave monotonicity, as the case may be. In other respects, the members of the population may have arbitrarily different response functions. The notation $y_j(\cdot)$ gives succinct and convenient expression to the idea that response functions may vary across the population.

Econometric analysis has a long tradition of expressing variation in treatment response in terms of variation in covariates. This complement interprets the assumption of monotone response from that perspective; semi-monotonicity and concave monotonicity can be interpreted similarly.

Let each person j have a covariate vector $u_j \in U$. These covariates may include the observable covariate x, but there is no need here to distinguish observable from unobservable covariates. A standard econometric response model expresses $y_j(\cdot)$ as

$$y_j(t) = y^*(t, u_j).\qquad (8B.1)$$

The function $y^*(\cdot, \cdot)$ mapping $T \times U$ into Y is common to all $j \in J$.

In terms of (8B.1), $y_j(t)$ is the outcome that person j would experience if he were to receive treatment t while holding his covariates fixed at the realized value u_j. Monotonicity of $y_j(\cdot)$ is equivalent to monotonicity of

$y^*(\cdot, u_j)$, with $t \geq s \Rightarrow y^*(t, u_j) \geq y^*(s, u_j)$. Random variable $y(t)$ expresses the outcomes that would be experienced if all members of the population were to receive treatment t while holding their covariates fixed at their realized values u_j, $j \in J$. Treatment effects $D[y(t)]-D[y(s)]$ and $D[y(t)-y(s)]$ compare the outcomes that would be experienced under treatments s and t if the covariates were held fixed at their realized values.

An alternative interpretation of $y_j(t)$ becomes available if we generalize the response model by supposing that variation in treatments induces variation in covariates. Let the *covariate response function* $u_j(\cdot)$: $T \to U$ map treatments into covariates, let $u_j \equiv u_j(z_j)$, and replace (8B.1) by

$$y_j(t) = y^*[t, u_j(t)]. \qquad (8B.2)$$

In this formulation, $y_j(t)$ is the outcome that person j would experience if he were to receive treatment t and his covariates were to take the value $u_j(t)$. Monotonicity of $y_j(\cdot)$ is equivalent to monotonicity of $y^*[\cdot, u_j(\cdot)]$ considered as a function of t. Treatment effects $D[y(t)]-D[y(s)]$ and $D[y(t)-y(s)]$ compare the outcomes experienced under treatments s and t taking account of induced variation in covariates.

The two interpretations of $y_j(t)$ are not contradictory. Propositions 8.1 through 8.3 apply to the thought experiment with covariates held fixed at realized values if $y^*(\cdot, u_j)$ is monotone on T. The propositions apply to the thought experiment with induced variation in covariates if $y^*[\cdot, u_j(\cdot)]$ is monotone on T. The propositions apply to both thought experiments if y^* is monotone in both senses. In this last case, one should not conclude that $y^*(t, u_j) = y^*[t, u_j(t)]$, but one can conclude that $y^*(t, u_j)$ and $y^*[t, u_j(t)]$ both lie within the common sharp bound $[y_{0j}(t), y_{1j}(t)]$.

Endnotes

Sources and Historical Notes

The analysis in this chapter originally appeared in Manski (1997a).

Text Notes

1. If T and Y do not have finite lower bounds, assuming that response is concave-monotone has no identifying power beyond assuming that response is monotone. Even assuming that response is linear-monotone has no additional identifying power. To see this, let $Y = R$ and suppose that

$$y_j(t) = \beta_j t + u_j,$$

where $\beta_j \geq 0$ is a person-specific slope parameter and u_j is a person-specific intercept. Observation of (y_j, z_j) reveals that $u_j = y_j - \beta_j z_j$, so

$$y_j(t) = \beta_j(t - z_j) + y_j.$$

For $s \in T$ and $t \in T$ with $t > s$, the sharp bound on $\{y_j(t), y_j(s)\}$ is

$$s < t < z_j \quad \Rightarrow \quad -\infty \leq y_j(s) \leq y_j(t) \leq y_j,$$

$$s < t = z_j \quad \Rightarrow \quad -\infty \leq y_j(s) \leq y_j(t) = y_j,$$

$$s < z_j < t \quad \Rightarrow \quad -\infty \leq y_j(s) \leq y_j \leq y_j(t) \leq \infty,$$

$$s = z_j < t \quad \Rightarrow \quad y_j = y_j(s) \leq y_j(t) \leq \infty,$$

$$z_j < s < t \quad \Rightarrow \quad y_j \leq y_j(s) \leq y_j(t) \leq \infty.$$

This is the same as the bound (8.12) obtained when it was assumed only that response is monotone. Hence, adding the linearity assumption leaves unchanged the conclusions to Propositions 8.1 through 8.3.

2. The early econometrics literature using instrumental variables to point-identify linear models of market demand (e.g., Wright, 1928; Reiersol, 1945) only assumed that u has zero covariance with v. However, the modern literature has generally maintained at least assumption (8A.2). Manski (1988, pp. 25–26 and Section 6.1) discusses the history and exposits the use of these assumptions.

9

Monotone Instrumental Variables

9.1. Equalities and Inequalities

To cope with the selection problem, researchers studying treatment response have long made use of distributional assumptions asserting forms of independence between outcomes and instrumental variables. Section 7.4 described the identifying power of various such assumptions. Complement 8A showed that point identification can be achieved by combining mean independence (assumption MI) with the assumption that all response functions are linear in treatments and have the same slope parameter.

Although independence assumptions are applied widely, their credibility in non-experimental settings often is a matter of considerable disagreement, with empirical researchers frequently debating whether some covariate is or is not a "valid instrument." There is therefore good reason to consider weaker assumptions that may be more credible. When the set V of instrumental-variable values is ordered, a simple way to weaken independence assumptions is to replace equalities with weak inequalities.

This chapter studies identification of mean treatment response when the equalities defining assumption MI are replaced with weak inequalities. Let t ∈ T. Assumption MI asserts that

$$E[y(t)|\ v] = E[y(t)].\qquad(9.1)$$

Replacement of the equality in (9.1) with a weak inequality yields the assumption of mean monotonicity (MM):

Assumption MM: Let V be an ordered set. Let $(v_1, v_2) \in V \times V$. Then

$$v_2 \geq v_1 \;\Rightarrow\; E[y(t)\,|\,v=v_2] \geq E[y(t)\,|\,v=v_1]. \qquad (9.2)$$

Assumption MM was introduced in Chapter 2 in the context of prediction with missing outcome data. Inequality (9.2) applies the assumption to the analysis of treatment response.

A particularly interesting case occurs when the instrumental variable is the realized treatment in the study population; that is, when $v = z$. Then assumption MI becomes the assumption of Means Missing at Random (MMAR):

$$E[y(t)\,|\,z] \;=\; E[y(t)]. \qquad (9.3)$$

Assumption MM becomes the assumption of *monotone treatment selection* (MTS):

Assumption MTS: Let T be an ordered set. Let $(t_1, t_2) \in T \times T$. Then

$$t_2 \geq t_1 \;\Rightarrow\; E[y(t)\,|\,z=t_2] \geq E[y(t)\,|\,z=t_1]. \qquad (9.4)$$

Assumption MTS was introduced in Chapter 2, in the context of prediction with missing outcome data, under the name *Means Missing Monotonically* (MMM). The name MTS is more descriptive in the present setting.

The Returns to Schooling

Assumption MM is of applied interest to the extent that it is more credible than assumption MI. Economic analysis of the returns to schooling illustrates the potential for gains in credibility.

Labor economists studying the returns to schooling usually suppose that each person j has a human-capital production function $y_j(t)$, giving the wage that j would receive were he to have t years of schooling. Observing realized wages and schooling, labor economists seek to learn about the population distribution of these production functions.

Empirical researchers often impose assumption MMAR. However, this assumption enjoys little credibility among economists. Perhaps the main reason is that various models of schooling choice and wage determination predict that persons with higher ability tend to have higher wage functions and tend to choose more schooling than do persons with lower ability. Assumption MMAR violates this prediction.

Assumption MTS is consistent with economic thinking about schooling choice and wage determination. This assumption asserts that persons who choose more schooling have weakly higher mean wage functions than do those who select less schooling. Thus, when studying the returns to schooling, assumption MTS is more credible than assumption MMAR.

Monotone Treatment Selection and Monotone Treatment Response

Assumption MTS is distinct from the assumption of monotone treatment response (MTR) studied in Chapter 8. Assumption MTR asserts that

$$t \geq s \; \Rightarrow \; y_j(t) \geq y_j(s) \qquad (9.5)$$

for all persons j and all treatment pairs (s, t). The inequalities in (9.4) and (9.5) express distinct properties of response functions. In principle, both assumptions could hold, or one, or neither.

To illustrate how Assumptions MTS and MTR differ, consider the variation of wages with schooling. Labor economists often say that "wages increase with schooling." Assumptions MTS and MTR interpret this statement in different ways. The MTS interpretation is that persons who select higher levels of schooling have weakly higher mean wage functions than do those who select lower levels of schooling. The MTR interpretation is that each person's wage function is weakly increasing in conjectured years of schooling. As discussed above, assumption MTS is consistent with economic models of schooling choice and wage determination that predict that persons with higher ability tend to have higher wage functions and tend to choose more schooling than do persons with lower ability. Assumption MTR expresses the standard economic view that education is a production process in which schooling is the input and wage is the output. Hence, wages increase with conjectured schooling.

Perhaps the most interesting finding in this chapter is that, when imposed together, Assumptions MTS and MTR have considerable identifying power. This is shown in Section 9.3. As prelude, Section 9.2 studies the identifying power of assumption MM alone, not combined with other assumptions.

The findings in this chapter apply immediately to sub-populations indexed by values of an observable covariate x. To simplify notation, the analysis does not explicitly condition on x.

9.2. Mean Monotonicity

Proposition 9.1 gives the identification region for $E[y(t)]$ under assumption MM. The proposition is an immediate extension of Proposition 2.6, so the proof is omitted.

Proposition 9.1: (a) Let V be an ordered set. Let assumption MM hold. Then the identification region for $E[y(t)]$ is the closed interval

$$H_{MM}\{E[y(t)]\} =$$

$$[\ \sum_{v \in V} P(v = v)(\max_{v' \le v} E\{y(t)\cdot 1[z = t] + y_0\cdot 1[z \ne t]\,|\,v = v'\}),$$

$$\sum_{v \in V} P(v = v)\,(\min_{v' \ge v} E\{y(t)\cdot 1[z = t] + y_1\cdot 1[z \ne t]\,|\,v = v'\})]. \quad (9.6)$$

(b) Let $H_{MM}\{E[y(t)]\}$ be empty. Then assumption MM does not hold. □

Suppose that the instrumental variable is the realized treatment z. Then Proposition 9.1 implies this identification region under assumption MTS:

Corollary 9.1.1: Let T be an ordered set. Let assumption MTS hold. Then the identification region for $E[y(t)]$ is the closed interval

$$H_{MTS}\{E[y(t)]\} =$$

$$[P(z < t)y_0 + P(z \ge t)E(y\,|\,z = t),\ P(z > t)y_1 + P(z \le t)E(y\,|\,z = t)]. \quad (9.7)$$
□

Proof: Application of Proposition 9.1 with $V = T$ and $v = z$ gives

$$H_{MTS}\{E[y(t)]\} =$$

$$[\ \sum_{s \in T} P(z = s)(\max_{s' \le s} E\{y(t)\cdot 1[z = t] + y_0\cdot 1[z \ne t]\,|\,z = s'\}),$$

$$\sum_{s \in T} P(z = s)(\min_{s' \ge s} E\{y(t)\cdot 1[z = t] + y_1\cdot 1[z \ne t]\,|\,z = s'\})]. \quad (9.8)$$

The lower endpoint in (9.8) reduces to the one in (9.7). To show this, observe that

$$s' < t \ \Rightarrow\ E\{y(t)\cdot 1[z = t] + y_0\cdot 1[z \ne t]\,|\,z = s'\} \ = \ y_0,$$

$$s' = t \ \Rightarrow\ E\{y(t)\cdot 1[z = t] + y_0\cdot 1[z \ne t]\,|\,z = s'\} \ = \ E(y\,|\,z = t),$$

$$s' > t \ \Rightarrow\ E\{y(t)\cdot 1[z = t] + y_0\cdot 1[z \ne t]\,|\,z = s'\} \ = \ y_0.$$

Hence

$$s < t \Rightarrow \max_{s' \le s} E\{y(t) \cdot 1[z = t] + y_0 \cdot 1[z \ne t] | z = s'\} = y_0,$$

$$s \ge t \Rightarrow \max_{s' \le s} E\{y(t) \cdot 1[z = t] + y_0 \cdot 1[z \ne t] | z = s'\} = E(y | z = t).$$

This yields the lower endpoint in (9.7). The proof for the upper endpoint is analogous.

Q. E. D.

It is revealing to compare Corollary 9.1.1 with the identification region for $E[y(t)]$ using the empirical evidence alone. The region using the empirical evidence alone is

$$H\{E[y(t)]\} = [P(z \ne t)y_0 + P(z = t)E(y | z = t), \; P(z \ne t)y_1 + P(z = t)E(y | z = t)].$$
$$(9.9)$$

The widths of intervals (9.7) and (9.9) are respectively

$$\|H_{MTS}\{E[y(t)]\}\| = [E(y | z = t) - y_0] P(z < t) + [y_1 - E(y | z = t)]P(z > t)$$
$$(9.10)$$

and

$$\|H\{E[y(t)]\}\| = (y_1 - y_0)P(z < t) + (y_1 - y_0)P(z > t). \qquad (9.11)$$

The former region is narrower than the latter one. For example, if $P(z < t) = P(z > t)$, the former region has one-half the width of the latter one.

9.3. Mean Monotonicity and Mean Treatment Response

In this section, Assumptions MM and MTR both hold. Proposition 9.2 gives the resulting sharp bound on $E[y(t)]$.

Proposition 9.2: Let V and T be ordered sets. Let Assumptions MM and MTR hold. Then

$$\sum_{v \in V} P(v = v) \left(\max_{v' \le v} E\{y \cdot 1[t \ge z] + y_0 \cdot 1[t < z] | v = v'\} \right)$$

$$\le E[y(t)] \le \qquad\qquad (9.12)$$

$$\sum_{v \in V} P(v = v) \ (\min_{v' \geq v} E\{y \cdot 1[t \leq z] + y_1 \cdot 1[t > z] | v = v'\}).$$

This bound is sharp.   □

Proof: Corollary 8.1.1 showed that, for each $v \in V$, assumption MTR yields this sharp bound on the conditional mean $E[y(t) | v = v]$:

$$E\{y \cdot 1[t \geq z] + y_0 \cdot 1[t < z] | v = v\} \ \leq \ E[y(t) | v = v]$$

$$\leq \ E\{y \cdot 1[t \leq z] + y_1 \cdot 1[t > z] | v = v\}. \qquad (9.13)$$

Assumption MM implies that, for all $(v_1, v_2) \in V \times V$,

$$v_1 \leq v \leq v_2 \ \Rightarrow \ E[y(t) | v = v_1] \ \leq \ E[y(t) | v = v] \ \leq \ E[y(t) | v = v_2]. \quad (9.14)$$

Combining (9.13) and (9.14) shows that $E[y(t) | v = v]$ is no smaller than the MTR lower bound on $E[y(t) | v = v_1]$ and no larger than the MTR upper bound on $E[y(t) | v = v_2]$. This holds for all $v_1 \leq v$ and all $v_2 \geq v$. There are no other restrictions on $E[y(t) | v = v]$. Thus, the sharp MM-MTR bound on $E[y(t) | v = v]$ is

$$\max_{v' \leq v} E\{y \cdot 1[t \geq z] + y_0 \cdot 1[t < z] | v = v'\} \ \leq \ E[y(t) | v = v]$$

$$\leq \ \min_{v' \geq v} E\{y \cdot 1[t \leq z] + y_1 \cdot 1[t > z] | v = v'\}. \qquad (9.15)$$

Now, consider the marginal mean $E[y(t)]$. The Law of Iterated Expectations gives

$$E[y(t)] = \sum_{v \in V} P(v = v) E[y(t) | v = v]. \qquad (9.16)$$

Inequality (9.15) shows that the sharp MM-MTR lower and upper bounds on $E[y(t) | v = v]$ are weakly increasing in v. Hence, the sharp joint lower (upper) bound on $\{E[y(t) | v = v], v \in V\}$ is obtained by setting each of the quantities $E[y(t) | v = v], v \in V$ at its lower (upper) bound in (9.15). Inserting these lower and upper bounds into the right side of (9.16) yields the result.
                                                                    Q. E. D.

The lower (upper) bound in Proposition 9.2 is informative only if y_0 (y_1) is finite. Suppose, however, that v is the realized treatment z, so assumption MM becomes assumption MTS. Then the bound turns out to be informative even if Y has infinite range. Corollary 9.2.1 gives the result.

Corollary 9.2.1: Let T be an ordered set. Let Assumptions MTS and MTR hold. Then

$$\sum_{s<t} E(y|z=s){\cdot}P(z=s) + E(y|z=t){\cdot}P(z\geq t) \;\leq\; E[y(t)]$$

$$\leq\; \sum_{s>t} E(y|z=s){\cdot}P(z=s) + E(y|z=t){\cdot}P(z\leq t). \qquad (9.17)$$

This bound is sharp.   □

Proof: Application of Proposition 9.2 with $V=T$ and $v=z$ gives

$$\sum_{s\in T} P(z=s)\,\big(\,\max_{s'\leq s}\; E\{y\cdot1[t\geq z]+y_0{\cdot}1[t<z]|z=s'\}\big)$$

$$\leq\; E[y(t)] \;\leq \qquad (9.18)$$

$$\sum_{s\in T} P(z=s)\,\big(\,\min_{s'\geq s} E\{y\cdot1[t\leq z]+y_1{\cdot}1[t>z]|z=s'\}\big).$$

The lower bound in (9.18) reduces to the one in (9.17). To show this, observe that

$$s'\leq t \;\Rightarrow\; E\{y{\cdot}1[t\geq z]+y_0{\cdot}1[t<z]|z=s'\} \;=\; E(y|z=s'),$$

$$s'>t \;\Rightarrow\; E\{y{\cdot}1[t\geq z]+y_0{\cdot}1[t<z]|z=s'\} \;=\; y_0.$$

Hence

$$s<t \;\Rightarrow\; \max_{s'\leq s} E\{y{\cdot}1[t\geq z]+y_0{\cdot}1[t<z]|z=s'\}$$

$$= \max_{s'\leq s} E(y|z=s') \;=\; E(y|z=s),$$

$$s\geq t \;\Rightarrow\; \max_{s'\leq s} E\{y{\cdot}1[t\geq z]+y_0{\cdot}1[t<z]|z=s'\}$$

$$= \max_{s'\leq t} E(y|z=s') \;=\; E(y|z=t).$$

The final equalities hold because, by Assumptions MTS and MTR,

$$s' \le s \;\Rightarrow\; E(y|z=s') = E[y(s')|z=s']$$

$$\le\; E[y(s)|z=s'] \;\le\; E[y(s)|z=s] = E(y|z=s). \qquad (9.19)$$

This yields the lower bound. The proof for the upper bound is analogous.

Q. E. D.

Inequality (9.19) suggests a test of assumption MTS-MTR. Under this joint hypothesis, $E(y|z=s)$ must be a weakly increasing function of s. Hence, the hypothesis is refuted if $E(y|z=s)$ is not weakly increasing in s. This test is a weakened version of the stochastic dominance test proposed in Section 8.3 for testing the joint hypothesis that treatment response is monotone and that z is statistically independent of $y(\cdot)$.

Bounds on Average Treatment Effects

Propositions 9.1 and 9.2 give sharp bounds on mean outcomes for specified treatments. Let s and t be two such treatments, with s < t. Often the object of interest is the average treatment effect $E[y(t)] - E[y(s)]$.

As usual, a lower (upper) bound on $E[y(t)] - E[y(s)]$ can be constructed by subtracting the lower (upper) bound on $E[y(t)]$ from the upper (lower) bound on $E[y(s)]$. When the construction is based on assumption MM alone, the resulting bound on $E[y(t)] - E[y(s)]$ is sharp. This follows from the fact that assumption MM imposes no joint restrictions on the response to different treatments.

Analysis of sharpness is generally complex when the construction is based on assumption MM-MTR but is possible in the special case when Assumptions MTS and MTR are combined. Then the upper bound on $E[y(t)] - E[y(s)]$ is

$$E[y(t)] - E[y(s)] \;\le\; \sum_{t'>t} E(y|z=t')\cdot P(z=t') + E(y|z=t)\cdot P(z \le t)$$

$$-\sum_{s'<s} E(y|z=s')\cdot P(z=s') - E(y|z=s)\cdot P(z \ge s). \qquad (9.20)$$

It follows from (9.19) that the right side of (9.20) is non-negative and no smaller than $E(y|z=t) - E(y|z=s)$, which is the value of $E[y(t)] - E[y(s)]$ under assumption MMAR. It is jointly feasible for $E[y(t)]$ to take its maximal value and $E[y(s)]$ its minimal value, so inequality (9.20) is sharp.

A lower bound on $E[y(t)] - E[y(s)]$ may be constructed in the same manner as (9.20), but the result is always nonpositive and is generically negative. Assumption MTR implies that $E[y(t)] - E[y(s)] \geq 0$, so the lower bound generically is not sharp.

9.4. Variations on the Theme

It is easy to think of variations on the theme of this chapter that warrant study and that may prove useful in empirical research. One would be to begin with assumption SI and weaken it to an assumption of stochastic dominance. Another would be to weaken assumption MI to some form of "approximate" mean independence. A way to formalize this would be to assert that, for $(v, v') \in V \times V$,

$$\|E[y(t)|v = v'] - E[y(t)|v = v]\| \leq C, \qquad (9.21)$$

where $C > 0$ is a specified constant. Yet another variation on the theme is to assert that a distributional assumption such as mean independence holds in part, but not all, of an observable population.[1]

Complement 9A. The Returns to Schooling

Manski and Pepper (2000) report an empirical analysis of the returns to schooling under Assumptions MTS and MTR. As indicated in Section 9.1, both assumptions are consistent with economic thinking about human capital accumulation. Even if these assumptions do not warrant unquestioned acceptance, they certainly merit serious consideration.

Data

The analysis used data from the National Longitudinal Survey of Youth (NLSY). In its base year of 1979, the NLSY interviewed 12,686 persons who were between the ages of 14 and 22 at that time. Nearly half of the respondents were randomly sampled, the remainder being selected to over-represent certain demographic groups. We restricted attention to the 1257 randomly sampled white males who reported in 1994 that they were full-time year-round workers with positive wages. The self-employed were excluded. Thus, the empirical analysis concerned the sub-population of persons who have the shared observable covariates

x = white males who reported in 1994 that they were full-time year-round workers but not self-employed, and who reported their wages.

The NLSY provides data on respondents' realized years of schooling and hourly wage in 1994. Thus z is realized years of schooling, the response variable $y_j(t)$ is the log(wage) that person j would experience if he were to have t years of schooling, and y_j is the observed hourly log(wage). The object of interest is the average treatment effect $\Delta(s, t) \equiv E[y(t)] - E[y(s)]$ for specified values of s and t.

(Note: Use of log(wage) rather than wage to measure the production of human capital follows the prevailing practice in labor economics. The reasons are not so much substantive as historical. Early researchers of the returns to schooling posed specific models for log(wages) that led to the establishment of research conventions followed by later researchers.)

Statistical Considerations

The bounds obtained in Propositions 9.1 and 9.2 are continuous functions of nonparametrically estimable conditional probabilities and mean responses. In this application, it was necessary only to estimate the MTS-MTR upper bound on $\Delta(s, t)$ given in (9.20). Thus, we had to estimate the probabilities $P(z)$ of realizing z years of schooling and the expectations $E(y|z)$ of log(wage) conditional on schooling. The empirical distribution of schooling was used to estimate $P(z)$ and the sample average log(wage) of respondents with z years of schooling was used to estimate $E(y|z = z)$. Estimation of the MTS-MTR upper bound was therefore a simple matter.

Asymptotically valid confidence intervals for the bounds may be computed using the delta method or bootstrap approaches. We applied the percentile bootstrap method. The bootstrap sampling distribution of an estimate of the MTS-MTR upper bound (9.20) is its sampling distribution under the assumption that the unknown distribution $P(y, z)$ equals the empirical distribution of these variables in the sample of 1257 randomly sampled NLSY respondents. The 0.95–quantile of the bootstrap sampling distribution is reported next to each upper-bound estimate.

Findings

Table 9.1 gives the estimates of $E(y|z)$ and $P(z)$ used to estimate the MTS-MTR bounds. The table shows that 41 percent of the NLSY respondents have 12 years of schooling and 19 percent have 16 years, but the support of the schooling distribution stretches from 8 years to 20 years. Hence, we were able to report findings on $\Delta(s, t)$ for t = 9 through 20 and $8 \le s < t$.

Section 9.3 showed that assumption MTS-MTR is a testable hypothesis, which should be rejected if $E(y|z = s)$ is not weakly increasing in s. The estimate of $E(y|z = s)$ in Table 9.1 for the most part does increase with s, but there are occasional dips. Computing a uniform 95 percent confidence band for the estimate of $E(y|z)$, we found that the band contains everywhere monotone functions. Hence, we proceeded on the basis that assumption MTS-MTR is consistent with the empirical evidence.

Table 9.1: Empirical Mean log(wage) and Distribution of Years of Schooling

z	$E(y|z)$	$P(z)$	Sample Size
8	2.249	0.014	18
9	2.302	0.018	22
10	2.195	0.018	23
11	2.346	0.025	32
12	2.496	0.413	519
13	2.658	0.074	93
14	2.639	0.083	104
15	2.693	0.035	44
16	2.870	0.189	238
17	2.775	0.038	48
18	3.006	0.051	64
19	3.009	0.020	25
20	2.936	0.021	27
Total		1	1257

Table 9.2 reports the estimates and bootstrap 0.95–quantiles of the MTS-MTR upper bounds on $\Delta(t-1, t)$, $t = 9, \ldots, 20$ followed by the upper bound on $\Delta(12, 16)$, which compares high school completion with college completion. Point estimates of these average treatment effects under assumption MMAR may be obtained directly from the first column of Table 9.1. Under this assumption, $\Delta(s, t) = E(y|z = t) - E(y|z = s)$.

To provide context for the results, it is useful to review the point estimates of $\Delta(t - 1, t)$ reported in the empirical literature on the returns to schooling. Most of the point estimates cited in the survey by Card (1994) are between 0.07 and 0.09. Card (1993) reports a point estimate of 0.132. Ashenfelter and Krueger (1994) report various estimates and conclude that (p. 1171): "our best estimate is that increased schooling increases average wage rates by about 12–16 percent per year completed."

Table 9.2: MTS-MTR Upper Bounds on Returns to Schooling

| | | Upper Bound on $\Delta(s, t)$ | |
s	t	Estimate	Bootstrap 0.95-Quantile
8	9	0.390	0.531
9	10	0.334	0.408
10	11	0.445	0.525
11	12	0.313	0.416
12	13	0.253	0.307
13	14	0.159	0.226
14	15	0.202	0.288
15	16	0.304	0.369
16	17	0.165	0.256
17	18	0.386	0.485
18	19	0.368	0.539
19	20	0.296	0.486
12	16	0.397	0.450

None of the estimates of upper bounds on $\Delta(t - 1, t)$ in Table 9.2 lies below the point estimates reported in the literature. The smallest of the upper-bound estimates are 0.159 for $\Delta(13, 14)$ and 0.165 for $\Delta(16, 17)$. These are about equal to the largest of the available point estimates, namely those in Ashenfelter and Krueger (1994). It may therefore appear that assumption MTS-MTR does not, in this application, have sufficient identifying power to affect current thinking about the magnitude of the returns to schooling.

A different conclusion emerges with consideration of the upper bound on $\Delta(12, 16)$. We estimate that completion of a four-year college yields at most an increase of 0.397 in mean log(wage) relative to completion of high school. This implies that the average value of the four year-by-year treatment effects $\Delta(12, 13)$, $\Delta(13, 14)$, $\Delta(14, 15)$, and $\Delta(15, 16)$ is at most 0.099, which is well below the point estimates of Card (1993) and Ashenfelter and Krueger (1994). This conclusion continues in force if, acting conservatively, one uses the bootstrap 0.95–quantile of 0.450 to estimate the upper bound on $\Delta(12, 16)$. Then the implied upper bound on the average value of the year-by-year treatment effects is 0.113. Thus, we found that, under assumption MTS-MTR, the returns to college-level schooling are smaller than some of the point estimates that have been reported.

Endnotes

Sources and Historical Notes

The analysis in this chapter originally appeared in Manski and Pepper (2000).

Text Notes

1. Hotz, Mullins, and Sanders (1997) study aspects of this last variation on the theme. They suppose that assumption MI holds in a population of interest. However, the observed population is a probability mixture of this population and another in which the assumption does not hold. Their analysis of *contaminated instruments* exploits the findings on contaminated sampling in Chapter 4.

10

The Mixing Problem

10.1. Within-Group Treatment Variation

A broad concern of the analysis of treatment response is extrapolation from one treatment rule to another. A planner or researcher observes the distribution $P(y, z, x)$ of (outcomes, treatments, covariates) realized under some *status quo* treatment rule and wants to learn the distribution of outcomes that would occur under a conjectural rule.

The planning problem of Chapter 7 motivated interest in predicting outcomes under rules in which treatment may vary across persons with different values of the observable covariate x but persons with the same value of x receive the same treatment. Chapters 8 and 9 continued to focus on rules that mandate uniform treatment of persons with the same observable covariates. Thus, these chapters studied identification of the outcome distributions $\{P[y(t)|x = x], t \in T, x \in X\}$ and of treatment effects that compare alternative mandated treatments.

Extrapolation from Randomized Experiments

This chapter studies prediction of outcomes when treatment may vary within the group of persons who share the same value of the covariates x. Within-group treatment variation may occur whenever treatment choices are made not by the planner of Chapter 7 but rather by other decision makers who can differentiate among persons with the same value of x. Within-group variation is particularly common when treatment choice is decentralized, each member of the population selecting his own treatment. For example, medical patients may choose among the several treatment options that physicians propose, youth may choose among a range of schooling alternatives, and so on.

This chapter specifically studies extrapolation from classical randomized experiments. As described in Chapter 7, the status quo rule in a classical experiment randomly places subjects in designated treatment groups, and all subjects comply with their designated treatments. A classical experiment credibly point-identifies outcome distributions under rules that mandate uniform treatment of persons with the same observable covariates, and enables a planner to make optimal treatment choices. A classical experiment does not point-identify outcome distributions under rules in which treatment may vary within groups. The task is to characterize what an experiment does reveal about outcomes under such treatment rules.

The Perry Preschool Project

An illustration helps to motivate the question under study and provides some insight into the underlying issues. A notable early use of experiments with random assignment of treatments to evaluate anti-poverty programs was the Perry Preschool Project begun in the early 1960s. Intensive educational and social services were provided to a random sample of about sixty black children, aged three and four, living in a low-income neighborhood of Ypsilanti, Michigan. No special services were provided to a second random sample of such children drawn to serve as a control group. The treatment and control groups were subsequently followed into adulthood. Among other things, it was found that 67 percent of the treatment group and 49 percent of the control group were high school graduates by age 19. This and similar findings for other outcomes have been cited widely as evidence that intensive early childhood educational interventions improve the outcomes of children at risk.[1]

Let $t = 1$ be the educational and social services provided to children participating in the project, and let $t = 0$ be the services available to children in the control group. Let $y(t)$ be a binary variable indicating high school graduation by age 19. For purposes of this illustration, consider the Perry Preschool Project to be a classical randomized experiment and ignore the fact that the sample sizes were on the small side. With these idealizations, the experimental data revealed that $P[y(1) = 1] = 0.67$ and $P[y(0) = 1] = 0.49$. Thus, the high school graduation probability would be 0.67 if all children in the relevant population were to receive the services provided by the Perry Preschool Project and would be 0.49 if none of them were to receive these services.

The question is this: What does the experiment reveal about the probability of high school graduation under a treatment rule in which some children receive the Perry Preschool services and the rest do not? For example, what would be the probability of graduation if budget limitations were to require rationing of services? What would it be if some parents were to refuse to allow their children to receive the services?

It might be conjectured that, if some children were to receive the Perry Preschool services and the rest were not, the high school graduation probability would necessarily lie between those observed in the Perry Preschool control and treatment groups, namely 0.49 and 0.67. This conjecture is correct under certain assumptions but not in general.

The experiment alone reveals only that the graduation rate would lie between 0.16 and 1. To see why, observe that each member of the population has one of these four values for $[y(1), y(0)]$:

$$[y(1) = 0, y(0) = 0], \qquad\qquad [y(1) = 0, y(0) = 1],$$
$$[y(1) = 1, y(0) = 0], \qquad\qquad [y(1) = 1, y(0) = 1].$$

Treatment assignment has no impact on persons for whom $y(1) = y(0)$ but determines the outcomes of persons for whom $y(1) \neq y(0)$. The highest feasible graduation rate is attained by a treatment rule that always selects the treatment with the better graduation outcome, and so gives treatment 1 to each person with $[y(1) = 1, y(0) = 0]$ and treatment 0 to each person with $[y(1) = 0, y(0) = 1]$. Then the only persons who do not graduate are those with $[y(1) = 0, y(0) = 0]$, so the graduation rate is $1 - P[y(1) = 0, y(0) = 0]$. Symmetrically, the lowest feasible graduation rate is attained by a rule that, by design or error, gives treatment 0 to each person with $[y(1) = 1, y(0) = 0]$ and treatment 1 to each person with $[y(1) = 0, y(0) = 1]$. Then the only persons who graduate are those with $[y(1) = 1, y(0) = 1]$, so the graduation rate is $P[y(1) = 1, y(0) = 1]$.

The experiment cannot reveal the joint probabilities $P[y(1) = 0, y(0) = 0]$ and $P[y(1) = 1, y(0) = 1]$ because treatments 1 and 0 are mutually exclusive. The experiment does reveal the marginal probabilities $P[y(1) = 1] = 0.67$ and $P[y(0) = 1] = 0.49$. It can be shown that among all joint distributions $P[y(1), y(0)]$ that are consistent with these marginals, there is one that minimizes both $P[y(1) = 0, y(0) = 0]$ and $P[y(1) = 1, y(0) = 1]$. This is

$$P[y(1) = 0, y(0) = 0] = 0, \qquad\qquad P[y(1) = 0, y(0) = 1] = 0.33,$$
$$P[y(1) = 1, y(0) = 0] = 0.51, \qquad\qquad P[y(1) = 1, y(0) = 1] = 0.16.$$

Hence, the highest graduation rate consistent with the experimental evidence is 1 and the lowest is 0.16.

From Marginals to Mixtures

Stripped to its essentials, extrapolation from a randomized experiment is a problem of inference on a probability mixture given knowledge of its marginals. This constitutes the *mixing problem*. The mixing problem should not be confused with the converse problem, studied in Chapters 4 and 5, in which one observes a probability mixture and wants to learn the distributions of the random variables that are mixed. Yet the two problems are related, as will become evident presently.

Let $\tau: J \rightarrow T$ be the treatment rule whose outcomes are to be predicted. Let

$$y(\tau) \equiv \sum_{t \in T} y(t) \cdot 1[\tau = t] \qquad (10.1)$$

be the random variable describing outcomes under rule τ. Thus $y(\tau)$ is a probability mixture of $[y(t), t \in T]$, whose distribution is

$$P[y(\tau)] = \sum_{t \in T} P[y(t)| \tau = t] \cdot P(\tau = t). \qquad (10.2)$$

A randomized experiment reveals the marginal outcome distributions $P[y(t)]$, $t \in T$. For each value of t, the Law of Total Probability gives

$$P[y(t)] = P[y(t)| \tau = t] \cdot P(\tau = t) + P[y(t)| \tau \neq t] \cdot P(\tau \neq t). \qquad (10.3)$$

Thus $P[y(t)]$ is the sum of $P[y(t)| \tau = t] \cdot P(\tau = t)$, which appears on the right side of (10.2), and $P[y(t)| \tau \neq t] \cdot P(\tau \neq t)$, which does not appear there.

The identification region for $P[y(\tau)]$ depends on what one knows about $P[y(\cdot)]$ and τ. Section 10.2 supposes that one knows the *treatment shares* $[P(\tau = t), t \in T]$ but has no other information. Section 10.3 studies extrapolation from the experiment alone.[2]

The findings in this chapter apply immediately to sub-populations indexed by values of an observable covariate x. To simplify notation, the analysis does not explicitly condition on x.

10.2. Known Treatment Shares

Analysis of identification with known treatment shares is the key step en route to study of extrapolation from the experiment alone. The case of known treatment shares is also of substantive interest. For example, resource constraints could limit implementation of the Perry Preschool intervention to part of the eligible population. Knowledge of the budget constraint and the cost per child of preschooling would suffice to determine the fraction of the population receiving the treatment. It may be more difficult to predict how school officials, social workers, and parents would interact to determine which children receive the treatment.

Suppose that the treatment shares under rule τ are known. Then equation (10.3) has the same structure as the contaminated sampling problem of Chapter 4. Let $p \equiv [P(\tau = t), t \in T]$ denote the vector of treatment shares under rule τ. For each $t \in T$, application of Proposition 4.1, part (a), gives the identification region for $P[y(t)|\tau = t]$:

$$H_p\{P[y(t)|\tau = t]\} \equiv \Gamma_Y \cap \{\{P[y(t)] - (1 - p_t)\gamma\}/p_t, \ \gamma \in \Gamma_Y\}. \qquad (10.4)$$

The experimental evidence for each treatment is uninformative about outcomes under other treatments. Hence, the joint identification region for $\{P[y(t)|\tau = t], t \in T\}$ is the Cartesian product $\times_{t \in T} H_p\{P[y(t)|\tau = t]\}$. Evaluating the right side of equation (10.2) at all feasible values of $\{P[y(t)|\tau = t], t \in T\}$ yields the identification region for $P[y(\tau)]$:

Proposition 10.1: Let $\{P[y(t)], t \in T\}$ and p be known. Then the identification region for $P[y(\tau)]$ is

$$H_p\{P[y(\tau)]\} \equiv \{\textstyle\sum_{t \in T} \eta_t \cdot p_t, \ \eta_t \in H_p\{P[y(t)|\tau = t]\}, t \in T\}. \qquad (10.5)$$

$\square$

Identification regions for event probabilities and for parameters that respect stochastic dominance similarly follow from Propositions 4.2 and 4.3. Corollaries 10.1.1 and 10.1.2 give the results.

Corollary 10.1.1: Let $B \subset Y$. Then the identification region for $P[y(\tau) \in B]$ is

$$H_p\{P[y(\tau) \in B]\} \equiv \{\textstyle\sum_{t \in T} \eta_t(B),$$

$$\eta_t(B) \in [\max \{0, P[y(t) \in B] - (1 - p_t)\}, \min\{p_t, P[y(t) \in B]\}], t \in T\}. \qquad (10.6)$$

$\square$

Proof: By (10.2),

$$P[y(\tau) \in B] = \textstyle\sum_{t \in T} P[y(t) \in B|\tau = t] \cdot p_t. \qquad (10.7)$$

Proposition 4.2, part (a), gives this identification region for $P[y(t) \in B|\tau = t]$:

$H_p\{P[y(t) \in B \mid \tau = t]\} \equiv$

$$[0, 1] \cap [\{P[y(t) \in B] - (1 - p_t)\}/p_t, \ P[y(t) \in B]/p_t]. \qquad (10.8)$$

Hence, the identification region for $P[y(t) \in B \mid \tau = t]p_t$ is

$H_p\{P[y(t) \in B \mid \tau = t]p_t\}$

$$\equiv [0, p_t] \cap [P[y(t) \in B] - (1 - p_t), P[y(t) \in B]] \qquad (10.9)$$

$$= [\max \{0, P[y(t) \in B] - (1 - p_t)\}, \ \min\{p_t, P[y(t) \in B]\}].$$

The identification region for $\{P[y(t) \in B \mid \tau = t]p_t, t \in T\}$ is the Cartesian product of the sets (10.9). Evaluating the right side of equation (10.7) at all feasible values of these quantities yields (10.6).

Q. E. D.

Corollary 10.1.2: Let Y be a subset of R that contains its lower and upper endpoints y_0 and y_1. For $t \in T$, define the distributions $L_{(p, t)}$ and $U_{(p, t)}$ on R as follows. For $r \in R$,

$$
\begin{aligned}
L_{(p, t)}[-\infty, r] &\equiv P[y(t) \le r]/p_t &\quad &\text{if } r < Q_{pt}[y(t)]\\
&\equiv 1 &\quad &\text{if } r \ge Q_{pt}[y(t)], &(10.10a)
\end{aligned}
$$

$$
\begin{aligned}
U_{(p, t)}[-\infty, t] &\equiv 0 &\quad &\text{if } r < Q_{1-pt}[y(t)]\\
&\equiv \{P[y(t) \le r] - (1 - p_t)\}/p_t &\quad &\text{if } r \ge Q_{1-pt}[y(t)]. &(10.10b)
\end{aligned}
$$

Let $D(\cdot)$ respect stochastic dominance. Then

$$D[\textstyle\sum_{t \in T} L_{(p, t)} \cdot p_t] \le D\{P[y(\tau)]\} \le D[\textstyle\sum_{t \in T} U_{(p, t)} \cdot p_t]. \qquad (10.11)$$

This bound is sharp.   □

Proof: The proof to Proposition 4.3, part (a) shows that $L_{(p, t)}$ and $U_{(p, t)}$ are the smallest and largest elements of identification region $H_p\{P[y(t) \mid \tau = t]\}$, with $L_{(p, t)}$ being stochastically dominated by every feasible distribution and $U_{(p, t)}$ stochastically dominating every such distribution. Hence, evaluating the right side of equation (10.2) at $(L_{(p, t)}, t \in T)$ and $(U_{(p, t)}, t \in T)$ yields the smallest and largest feasible values for any parameter of $P[y(\tau)]$ that respects stochastic dominance.

Q. E. D.

10.3. Extrapolation from the Experiment Alone

Now suppose that the treatment shares are unknown, so the only available information is the empirical evidence from the randomized experiment. Let S denote the unit simplex in $R^{|T|}$. The treatment shares can take any value in S. Hence, the identification region for $P[y(\tau)]$ using the empirical evidence alone is the union of the sets $H_p\{P[y(\tau)]\}$ across all $p \in S$. Analogous findings hold for event probabilities and for parameters that respect stochastic dominance. These findings are collected in Proposition 10.2.

Proposition 10.2: Let $\{P[y(t)], t \in T\}$ be known. Then the identification region for $P[y(\tau)]$ is

$$H\{P[y(\tau)]\} \;=\; \bigcup_{p \in S} H_p\{P[y(\tau)]\}. \tag{10.12}$$

Let $B \subset Y$. Then the identification region for $P[y(\tau) \in B]$ is

$$H\{P[y(\tau) \in B]\} \;=\; \bigcup_{p \in S} H_p\{P[y(\tau) \in B]\}. \tag{10.13}$$

Let $D(\cdot)$ respect stochastic dominance. Then

$$\inf_{p \in S} D[\textstyle\sum_{t \in T} L_{(p,t)} \cdot p_t] \;\leq\; D\{P[y(\tau)]\} \;\leq\; \sup_{p \in S} D[\textstyle\sum_{t \in T} U_{(p,t)} \cdot p_t]. \tag{10.14}$$

This bound is sharp.   □

   Proposition 10.2 is general but abstract. Consideration of the special case of identification of event probabilities when there are two treatments yields a result that is easy to grasp and apply. Corollary 10.2.1 gives this result.

Corollary 10.2.1: Let $T = \{0, 1\}$. Define $C \equiv P[y(1) \in B] + P[y(0) \in B]$. Then the identification region for $P[y(\tau) \in B]$ is

$$H\{P[y(\tau) \in B]\} \;=\; [\max(0, C - 1), \; \min(C, 1)]. \tag{10.15}$$
□

Proof: The corollary can be proved directly, albeit laboriously, by evaluation of equation (10.13) when T contains two elements. The derivation below uses a simple direct argument to show that the endpoints of the interval in (10.15) are sharp bounds on $P[y(\tau) \in B]$. This derivation formalizes the reasoning in the Perry Preschool illustration of Section 10.1.

The largest possible value of $P[y(\tau) \in B]$ is $1 - P[y(1) \notin B \cap y(0) \notin B]$. This is achieved by a rule that always chooses a treatment whose outcome lies in B, when such a treatment exists. The smallest possible value of $P[y(\tau) \in B]$ is $P[y(1) \in B \cap y(0) \in B]$. This is achieved by a rule that always chooses a treatment whose outcome lies in the complement of B, when such a treatment exists. Thus $P[y(\tau) \in B]$ must lie in the interval

$$P[y(1) \in B \cap y(0) \in B] \leq P[y(\tau) \in B] \leq 1 - P[y(1) \notin B \cap y(0) \notin B].$$
$$(10.16)$$

If $P[y(\cdot)]$ were known, (10.16) would be the sharp bound on $P[y(\tau) \in B]$. We are concerned, however, with the situation in which only the marginals $P[y(1)]$ and $P[y(0)]$ are known. In this situation, the sharp lower bound on $P[y(\tau) \in B]$ is the smallest value of $P[y(1) \in B \cap y(0) \in B]$ that is consistent with the known $P[y(1)]$ and $P[y(0)]$. The sharp upper bound is one minus the smallest feasible value of $P[y(1) \notin B \cap y(0) \notin B]$.

Let $A \subset Y$. It can be shown that knowledge of the marginals $P[y(1)]$ and $P[y(0)]$ implies this sharp bound on $P[y(1) \in A \cap y(0) \in A]$: [3]

$$\max\{0, P[y(1) \in A] + P[y(0) \in A] - 1\} \leq P[y(1) \in A \cap y(0) \in A]$$

$$\leq \min\{P[y(1) \in A], P[y(0) \in A]\}. \qquad (10.17)$$

Application of (10.17) with $A = B$ yields the lower bound on $P[y(\tau) \in B]$ in (10.15). Application of (10.17) with $A = Y - B$ yields the upper bound.
Q. E. D.

Observe that the identification region for $P[y(\tau) \in B]$ is informative from the left or the right but not from both sides simultaneously. The width of the region narrows toward 0 as C approaches 0 or 2 but widens toward 1 as C approaches 1. Thus, knowledge of the marginals may reveal a lot or a little about the magnitude of $P[y(\tau) \in B]$, depending on the empirical value of C. In the Perry Preschool illustration, $B = \{1\}$, $P[y(0) \in B] = 0.49$, $P[y(1) \in B] = 0.67$, and $C = 1.16$.

Complement 10A. Experiments Without Covariate Data

An interesting manifestation of the mixing problem occurs when a planner observes the treatments and outcomes realized in a classical randomized experiment but does not observe covariates of the experimental subjects. This informational situation is common in medical settings. Physicians often have extensive covariate information — medical histories, diagnostic test findings, and demographic attributes — for the patients that they treat. Physicians also often know the outcomes of randomized clinical trials evaluating alternative treatments. However, the medical journal articles that report the findings of clinical trials do not usually report much covariate information for the subjects of the experiment. Articles reporting clinical trials usually describe outcomes only for broad risk-factor groups.

To grasp the essence of the planner's problem, it suffices to consider the simplest non-trivial setting: that in which treatments, outcomes, and covariates are all binary. Thus, suppose that there are two treatments, say $t = 0$ and $t = 1$. The outcome $y(t)$ is binary, with values $y(t) = 0$ and $y(t) = 1$; hence $E[y(t)|x] = P[y(t) = 1|x]$. The covariate x is also binary, taking the values $x = a$ and $x = b$.

Even in this simple setting, analysis of the planner's problem turns out to be complex. There are four feasible treatment rules. These rules and their mean outcomes are

Treatment Rule τ (0, 0): All persons receive $t = 0$. The mean outcome is
$$M(0, 0) \equiv P[y(0) = 1].$$

Treatment Rule τ (1, 1): All persons receive $t = 1$. The mean outcome is
$$M(1, 1) \equiv P[y(1) = 1].$$

Treatment Rule τ (0, 1): Persons with $x = a$ receive $t = 0$, and persons with $x = b$ receive $t = 1$. The mean outcome is
$$M(0, 1) \equiv P[y(0) = 1|x = a] \cdot P(x = a) + P[y(1) = 1|x = b] \cdot P(x = b).$$

Treatment Rule τ (1, 0): Persons with $x = a$ receive $t = 1$, and persons with $x = b$ receive $t = 0$. The mean outcome is
$$M(1, 0) \equiv P[y(1) = 1|x = a] \cdot P(x = a) + P[y(0) = 1|x = b] \cdot P(x = b).$$

The Dominated Treatment Rules

Which of the four feasible treatment rules are dominated? The experiment reveals $M(0, 0)$ and $M(1, 1)$. Thus, rule τ (0, 0) is dominated if $M(0, 0) < M(1, 1)$, and rule τ (1, 1) is dominated if $M(1, 1) < M(0, 0)$. The planner is indifferent between these two rules if $M(0, 0) = M(1, 1)$.

The experiment does not reveal $M(0, 1)$ and $M(1, 0)$. However, Corollary 10.1.1 shows that the experiment in the study population and knowledge of the covariate distribution in the treatment population imply sharp bounds on these quantities. The sharp bounds on $M(0, 1)$ and $M(1, 0)$ are

$\max\{0, P[y(1) = 1] - P(x = a)\}$ + $\max\{0, P[y(0) = 1] - P(x = b)\}$

$\leq M(0, 1) \leq \min\{P(x = b), P[y(1) = 1]\} + \min\{P(x = a), P[y(0) = 1]\}$,

$\max\{0, P[y(1) = 1] - P(x = b)\}$ + $\max\{0, P[y(0) = 1] - P(x = a)\}$

$\leq M(1, 0) \leq \min\{P(x = a), P[y(1) = 1]\} + \min\{P(x = b), P[y(0) = 1]\}$.

The form of these bounds determines which treatment rules are dominated. Suppose that $P[y(0) = 1] \leq P[y(1) = 1]$ and $P(x = a) \leq P(x = b)$. Then rule $\tau(0, 0)$ is dominated by $\tau(1, 1)$. The dominance relations among the other rules depend on the ordering of $P[y(0) = 1]$, $P[y(1) = 1]$, $P(x = a)$, and $P(x = b)$. There are six distinct orderings to be considered:

Case 1: $P[y(0) = 1] \leq P[y(1) = 1] \leq P(x = a) \leq P(x = b)$.
  $0 \leq M(0, 1) \leq P[y(1) = 1] + P[y(0) = 1]$.
  $0 \leq M(1, 0) \leq P[y(1) = 1] + P[y(0) = 1]$.
Then rules $\tau(0, 1)$, $\tau(1, 0)$, and $\tau(1, 1)$ are undominated.

Case 2: $P[y(0) = 1] \leq P(x = a) \leq P[y(1) = 1] \leq P(x = b)$.
  $P[y(1) = 1] - P(x = a) \leq M(0, 1) \leq P[y(1) = 1] + P[y(0) = 1]$.
  $0 \leq M(1, 0) \leq P(x = a) + P[y(0) = 1]$.
Then rules $\tau(0, 1)$ and $\tau(1, 1)$ are undominated. Rule $\tau(1, 0)$ is dominated by rule $\tau(1, 1)$ if $P(x = a) + P[y(0) = 1] < P[y(1) = 1]$.

Case 3: $P[y(0) = 1] \leq P(x = a) \leq P(x = b) \leq P[y(1) = 1]$.
  $P[y(1) = 1] - P(x = a) \leq M(0, 1) \leq P(x = b) + P[y(0) = 1]$.
  $P[y(1) = 1] - P(x = b) \leq M(1, 0) \leq P(x = a) + P[y(0) = 1]$.
Then rule $\tau(1, 1)$ is undominated. Rule $\tau(0, 1)$ is dominated by rule $\tau(1, 1)$ if $P(x = b) + P[y(0) = 1] < P[y(1) = 1]$. Rule $\tau(1, 0)$ is dominated by rule $\tau(1, 1)$ if $P(x = a) + P[y(0) = 1] < P[y(1) = 1]$.

Case 4: $P(x = a) \leq P[y(0) = 1] \leq P[y(1) = 1] \leq P(x = b)$.
  $P[y(1) = 1] - P(x = a) \leq M(0, 1) \leq P[y(1) = 1] + P(x = a)$.
  $P[y(0) = 1] - P(x = a) \leq M(1, 0) \leq P(x = a) + P[y(0) = 1]$.
Then rules $\tau(1, 1)$ and $\tau(0, 1)$ are undominated. Rule $\tau(1, 0)$ is dominated by rule $\tau(1, 1)$ if $P(x = a) + P[y(0) = 1] < P[y(1) = 1]$.

Case 5: $P(x = a) \leq P[y(0) = 1] \leq P(x = b) \leq P[y(1) = 1]$.
  $P[y(1) = 1] - P(x = a) \leq M(0, 1) \leq 1$.
  $P[y(1) = 1] + P[y(0) = 1] - 1 \leq M(1, 0) \leq P(x = a) + P[y(0) = 1]$.
Then rules $\tau(1, 1)$ and $\tau(0, 1)$ are undominated. Rule $\tau(1, 0)$ is dominated

by rule $\tau(1, 1)$ if $P(x = a) + P[y(0) = 1] < P[y(1) = 1]$.

Case 6: $P(x = a) \le P(x = b) \le P[y(0) = 1] \le P[y(1) = 1]$.
          $P[y(1) = 1] + P[y(0) = 1] - 1 \le M(0, 1) \le 1$.
          $P[y(1) = 1] + P[y(0) = 1] - 1 \le M(1, 0) \le 1$.
Then rules $\tau(0, 1)$, $\tau(1, 0)$, and $\tau(1, 1)$ are undominated.

Cases 1 through 6 show that as many as three or as few as zero treatment rules are dominated, depending on the empirical values of $P[y(0) = 1]$, $P[y(1) = 1]$, $P(x = a)$, and $P(x = b)$. The one constancy is that rule $\tau(1, 1)$ is always undominated. Indeed, $\tau(1, 1)$ is always the maximin rule.

The Perry Preschool Project Revisited

To illustrate, consider the situation of a planner, perhaps a social worker, who is charged with making preschool treatment choices for low-income black children in Ypsilanti and whose objective is to maximize the high school graduation rate. The planner can assign each child to the Perry Preschool treatment or not. Suppose that the planner observes a binary covariate that describes each member of the population. For the sake of concreteness, let the covariate indicate the child's family status, with $x = a$ if the child has an intact two-parent family and $x = b$ otherwise.

The available outcome data reveal that rule $\tau(0, 0)$, where no children receive the Perry Preschool treatment, is dominated by rule $\tau(1, 1)$, where all children receive preschooling. The conclusions that the planner can draw about rules $\tau(0, 1)$ and $\tau(1, 0)$ depend on the covariate distribution $P(x)$.

Suppose that half the children have intact families, so $P(x = a) = P(x=b) = 0.5$. Then Case 3 holds. The bounds on mean outcomes under rules $\tau(0, 1)$ and $\tau(1, 0)$ are

$$0.17 \le M(0, 1) \le 0.99 \qquad 0.17 \le M(1, 0) \le 0.99.$$

These bounds imply that rules $\tau(0, 1)$ and $\tau(1, 0)$, which reverse one another's treatment assignments, have an enormously wide range of potential consequences for high school graduation. The best case for $\tau(0, 1)$ and the worst for $\tau(1, 0)$ both occur if the (unknown) graduation probabilities conditional on covariates are

$$P[y(0) = 1 \,|\, x = a] = 0.98, \qquad P[y(1) = 1 \,|\, x = a] = 0.34,$$
$$P[y(0) = 1 \,|\, x = b] = 0, \qquad P[y(1) = 1 \,|\, x = b] = 1.$$

These graduation probabilities, which yield $M(0, 1) = 0.99$ and $M(1, 0)$

= 0.17, are consistent with the experimental evidence that $P[y(0) = 1]$ = 0.49 and $P[y(1) = 1]$ = 0.67. They describe a possible world in which preschooling is necessary and sufficient for children in non-intact families to complete high school but substantially hurts the graduation prospects of children in intact families. There is another possible world with the reverse graduation probabilities: one in which $M(0, 1)$ = 0.17 and $M(1, 0)$ = 0.99. Hence, rules $\tau(0, 1)$, $\tau(1, 0)$, and $\tau(1, 1)$ are all undominated.

The planner faces a much less ambiguous choice problem if most children have non-intact families. Suppose that $P(x = a)$ = 0.1 and $P(x = b)$ = 0.9. Then Case 4 holds. The bounds on mean outcomes under rules $\tau(0, 1)$ and $\tau(1, 0)$ are

$$0.57 \leq M(0, 1) \leq 0.77 \qquad 0.39 \leq M(1, 0) \leq 0.59.$$

These bounds are much narrower than those obtained when half of all children have non-intact families. The upper bound on $M(1, 0)$ is 0.59, which is less than the known value of $M(1, 1)$, namely 0.67. Hence, rule $\tau(1, 0)$ is dominated. Recall that rule $\tau(0, 0)$ is also dominated. Thus, although the planner does not observe graduation probabilities conditional on covariates, he can nevertheless conclude that the 90 percent of children who have non-intact families should receive preschooling. The only ambiguity about treatment choice concerns the 10 percent of children who have intact families. Treatment rules $\tau(0, 1)$ and $\tau(1, 1)$ are undominated. Thus, in the absence of other information, the planner cannot determine whether children in intact families should or should not receive preschooling.

Endnotes

Sources and Historical Notes

This chapter extends analysis that originally appeared in Manski (1995, 1997b). Complement 10A is taken from Manski (2000).

Text Notes

1. See Berrueta-Clement et al. (1984) and Holden (1990).

2. Additional findings for other informational settings are reported in Manski (1997b) and Pepper (2003).

3. This was proved by Frechét (1951); see Ord (1972) for an exposition and Ruschendorf (1981) for a thorough analysis. It is elementary to show that $P[y(1) \in A \cap y(0) \in A]$ must lie within the bound. The upper bound holds because the event $[y(1) \in A \cap y(0) \in A]$ implies each of its component events $[y(1) \in A]$ and $[y(0) \in A]$. The lower bound holds because

$$1 \geq P[y(1) \in A \cup y(0) \in A]$$

$$= P[y(1) \in A] + P[y(0) \in A] - P[y(1) \in A \cap y(0) \in A].$$

Frechét's general analysis of the problem of inference on a joint distribution from knowledge of its marginals showed that bound (10.17) is sharp.

References

Angrist, J., G. Imbens, and D. Rubin (1996), "Identification of Causal Effects Using Instrumental Variables," *Journal of the American Statistical Association*, 91, 444–455.

Arabmazar, A. and P. Schmidt (1982), "An Investigation of the Robustness of the Tobit Estimator to Non-Normality," *Econometrica*, 50, 1055–1063.

Ashenfelter, O. and A. Krueger (1994), "Estimates of the Economic Returns to Schooling from a New Sample of Twins," *American Economic Review*, 84, 1157–1173.

Balke, A. and J. Pearl (1997), "Bounds on Treatment Effects from Studies with Imperfect Compliance," *Journal of the American Statistical Association*, 92, 1171–1177.

Bedford, T. and I. Meilijson (1997), "A Characterization of Marginal Distributions of (Possibly Dependent) Lifetime Variables which Right Censor Each Other," *The Annals of Statistics*, 25, 1622–1645.

Berger, J. (1985), *Statistical Decision Theory and Bayesian Analysis*, New York: Springer-Verlag.

Berkson, J. (1958), "Smoking and Lung Cancer: Some Observations on Two Recent Reports," *Journal of the American Statistical Association*, 53, 28–38.

Berrueta-Clement, J., L. Schweinhart, W. Barnett, A. Epstein, and D. Weikart (1984), *Changed Lives: The Effects of the Perry Preschool Program on Youths Through Age 19*, Ypsilanti, MI: High/Scope Press.

Brøndsted, A. (1983), *An Introduction to Convex Polytopes*, New York: Springer-Verlag.

Campbell, D. (1984), "Can We Be Scientific in Applied Social Science?," *Evaluation Studies Review Annual*, 9, 26–48.

167

Campbell, D. and R. Stanley (1963), *Experimental and Quasi-Experimental Designs for Research*, Chicago: Rand McNally.

Card, D. (1993), "Using Geographic Variation in College Proximity to Estimate the Return to Schooling," Working Paper 4483, Cambridge, MA: National Bureau of Economic Research.

Card, D. (1994), "Earnings, Schooling, and Ability Revisited," Working Paper 4832, Cambridge, MA: National Bureau of Economic Research.

Center for Human Resource Research (1992), *NLS Handbook 1992. The National Longitudinal Surveys of Labor Market Experience*, Columbus, OH: The Ohio State University.

Cochran, W. (1977), *Sampling Techniques*, Third Edition. New York: Wiley.

Cochran, W., F. Mosteller, and J. Tukey (1954), *Statistical Problems of the Kinsey Report on Sexual Behavior in the Human Male*, Washington, DC: American Statistical Association.

Cornfield, J. (1951), "A Method of Estimating Comparative Rates from Clinical Data. Applications to Cancer of the Lung, Breast, and Cervix," *Journal of the National Cancer Institute*, 11, 1269–1275.

Crowder, M. (1991), "On the Identifiability Crisis in Competing Risks Analysis," *Scandinavian Journal of Statistics*, 18, 223–233.

Cross, P. and C. Manski (2002), "Regressions, Short and Long," *Econometrica*, 70, 357–368.

Duncan, O. and B. Davis (1953), "An Alternative to Ecological Correlation," *American Sociological Review*, 18, 665–666.

Ellsberg, D. (1961), "Risk, Ambiguity, and the Savage Axioms," *Quarterly Journal of Economics*, 75, 643–669.

Fitzgerald, J., P. Gottschalk, and R. Moffitt (1998), "An Analysis of Sample Attrition in Panel Data," *Journal of Human Resources*, 33, 251–299.

Fleiss, J. (1981), *Statistical Methods for Rates and Proportions*, New York: Wiley.

Frechét, M. (1951), "Sur Les Tableaux de Correlation Donte les Marges sont Donnees," *Annals de l'Universite de Lyon A*, Series 3, 14, 53-77.

Freedman, D., S. Klein, M. Ostland, and M. Roberts (1998), "Review of *A Solution to the Ecological Inference Problem*, by G. King," *Journal of the American Statistical Association*, 93, 1518–1522.

Freedman, D., S. Klein, M. Ostland, and M. Roberts (1999), "Response to King's Comment," *Journal of the American Statistical Association*, 94, 355–357.

Freis, E.D., Materson, B.J., and Flamenbaum, W. (1983), "Comparison of Propranolol or Hydrochlorothiazide Alone for Treatment of Hypertension, III: Evaluation of the Renin-Angiotensin System," *The American Journal of Medicine*, 74, 1029–1041.

Frisch, R. (1934), *Statistical Confluence Analysis by Means of Complete Regression Systems*, Oslo, Norway: University Institute for Economics.

Goldberger, A. (1972), "Structural Equation Methods in the Social Sciences," *Econometrica*, 40, 979–1001.

Goldberger, A. (1983), "Abnormal Selection Bias," in T. Amemiya and I. Olkin (eds.), *Studies in Econometrics, Time Series, and Multivariate Statistics*, Orlando: Academic Press.

Goldberger, A. (1984), "Reverse Regression and Salary Discrimination," *Journal of Human Resources*, 19, 293-318.

Goldberger, A. (1991), *A Course in Econometrics*, Cambridge, MA: Harvard University Press.

Goodman, L. (1953), "Ecological Regressions and Behavior of Individuals," *American Sociological Review*, 18, 663–664.

Gronau, R. (1974), "Wage Comparisons–a Selectivity Bias," *Journal of Political Economy*, 82, 1119–1143.

Hampel, F., E. Ronchetti, P. Rousseeuw, and W. Stahel (1986), *Robust Statistics*, New York: Wiley.

Heckman, J. (1976), "The Common Structure of Statistical Models of Truncation, Sample Selection, and Limited Dependent Variables and a Simple Estimator for Such Models," *Annals of Economic and Social Measurement*, 5, 479–492.

Heckman, J., J. Smith, and N. Clements (1997), "Making the Most out of Programme Evaluations and Social Experiments: Accounting for Heterogeneity in Programme Impacts," *Review of Economic Studies*, 64, 487–535.

Hirano, K., G. Imbens, G. Ridder, and D. Rubin (2001), "Combining Panel Data Sets with Attrition and Refreshment Samples," *Econometrica*, 69, 1645–1659.

Holden, C. (1990), "Head Start Enters Adulthood," *Science*, 247, 1400–1402.

Hood, W. and T. Koopmans (eds.) (1953), *Studies in Econometric Method*, New York: Wiley.

Horowitz, J. and C. Manski (1995), "Identification and Robustness with Contaminated and Corrupted Data," *Econometrica*, 63, 281–302.

Horowitz, J. and C. Manski (1997), "What Can Be Learned About Population Parameters when the Data Are Contaminated?," in C. R. Rao and G. S. Maddala (eds.), *Handbook of Statistics*, Vol. 15: Robust Statistics, Amsterdam: North-Holland, pp.439–466.

Horowitz, J. and C. Manski (1998), "Censoring of Outcomes and Regressors due to Survey Nonresponse: Identification and Estimation Using Weights and Imputations," *Journal of Econometrics*, 84, 37–58.

Horowitz, J. and C. Manski (2000), "Nonparametric Analysis of Randomized Experiments with Missing Covariate and Outcome Data," *Journal of the American Statistical Association*, 95, 77–84.

Horowitz, J. and C. Manski (2001), "Imprecise Identification from Incomplete Data," *Proceedings of the 2nd International Symposium on Imprecise Probabilities and Their Applications*, http://ippserv.rug.ac.be/~isipta01/proceedings/index.html.

Hotz, J., C. Mullins, and S. Sanders (1997), "Bounding Causal Effects Using Data from a Contaminated Natural Experiment: Analyzing the Effects of Teenage Childbearing," *Review of Economic Studies*, 64, 575–603.

Hsieh, D., C. Manski, and D. McFadden (1985), "Estimation of Response Probabilities from Augmented Retrospective Observations," *Journal of the American Statistical Association*, 80, 651-662.

Huber, P. (1964), "Robust Estimation of a Location Parameter," *Annals of Mathematical Statistics*, 35, 73–101.

Huber, P. (1981), *Robust Statistics*, New York: Wiley.

Hurd, M. (1979), "Estimation in Truncated Samples when There Is Heteroskedasticity," *Journal of Econometrics*, 11, 247–258.

Imbens, G. and J. Angrist (1994), "Identification and Estimation of Local Average Treatment Effects," *Econometrica*, 62, 467–476.

Keynes, J. (1921), *A Treatise on Probability*, London: MacMillan.

King, G. (1997), *A Solution to the Ecological Inference Problem: Reconstructing Individual Behavior from Aggregate Data*, Princeton: Princeton University Press.

King, G. (1999), "The Future of Ecological Inference Research: A Comment on Freedman et al.," *Journal of the American Statistical Association*, 94, 352–355.

King, G. and L. Zeng (2002), "Estimating Risk and Rate Levels, Ratios and Differences in Case-Control Studies," *Statistics in Medicine*, 21, 1409–1427.

Klepper, S. and E. Leamer (1984), "Consistent Sets of Estimates for Regressions with Errors in All Variables," *Econometrica*, 52, 163–183.

Knight, F. (1921), *Risk, Uncertainty, and Profit*, Boston: Houghton-Mifflin.

Koopmans, T. (1949), "Identification Problems in Economic Model Construction," *Econometrica*, 17, 125–144.

Lindley, D. and M. Novick (1981), "The Role of Exchangeability in Inference," *Annals of Statistics*, 9, 45–58.

Little, R. (1992), "Regression with Missing X's: A Review," *Journal of the American Statistical Association*, 87, 1227–1237.

Little, R. and D. Rubin (1987), *Statistical Analysis with Missing Data*, New York: Wiley.

Maddala, G. S. (1983), *Limited-Dependent and Qualitative Variables in Econometrics*, Cambridge, UK: Cambridge University Press.

Manski, C. (1988), *Analog Estimation Methods in Econometrics*, London: Chapman & Hall.

Manski, C. (1989), "Anatomy of the Selection Problem," *Journal of Human Resources*, 24, 343–360.

Manski, C. (1990), "Nonparametric Bounds on Treatment Effects," *American Economic Review Papers and Proceedings*, 80, 319–323.

Manski, C. (1994), "The Selection Problem," in C. Sims (ed.), *Advances in Econometrics, Sixth World Congress*, Cambridge, UK: Cambridge University Press, pp.143–170.

Manski, C. (1995), *Identification Problems in the Social Sciences*, Cambridge, MA: Harvard University Press.

Manski, C. (1997a), "Monotone Treatment Response," *Econometrica*, 65, 1311–1334.

Manski, C. (1997b), "The Mixing Problem in Programme Evaluation," *Review of Economic Studies*, 64, 537–553.

Manski, C. (2000), "Identification Problems and Decisions Under Ambiguity: Empirical Analysis of Treatment Response and Normative Analysis of Treatment Choice," *Journal of Econometrics*, 95, 415–442.

Manski, C. (2001), "Nonparametric Identification Under Response-Based Sampling," in C. Hsiao, K. Morimune, and J. Powell (eds.), *Nonlinear Statistical Inference: Essays in Honor of Takeshi Amemiya*, New York: Cambridge University Press.

Manski, C. (2002), "Treatment Choice Under Ambiguity Induced by Inferential Problems," *Journal of Statistical Planning and Inference*, 105, 67–82.

Manski, C. (2003), "Social Learning from Private Experiences: The Dynamics of the Selection Problem," *Review of Economic Studies*, forthcoming.

Manski, C. and S. Lerman (1977), "The Estimation of Choice Probabilities from Choice-Based Samples," *Econometrica*, 45, 1977-1988.

Manski, C. and D. Nagin (1998), "Bounding Disagreements About Treatment Effects: A Case Study of Sentencing and Recidivism," *Sociological Methodology*, 28, 99–137.

Manski, C. and J. Pepper (2000), "Monotone Instrumental Variables: With an Application to the Returns to Schooling," *Econometrica*, 68, 997–1010.

Manski, C. and E. Tamer (2002), "Inference on Regressions with Interval Data on a Regressor or Outcome," *Econometrica*, 70, 519–546.

Materson, B., D. Reda, and W. Cushman (1995), "Department of Veterans Affairs Single-Drug Therapy of Hypertension Study: Revised Figures and New Data," *American Journal of Hypertension*, 8, 189–192.

Materson, B., D. Reda, W. Cushman, B. Massie, E. Freis, M. Kochar, R. Hamburger, C. Fye, R. Lakshman, J. Gottdiener, E. Ramirez, and W. Henderson (1993), "Single-Drug Therapy for Hypertension in Men: A Comparison of Six Antihypertensive Agents with Placebo," *The New England Journal of Medicine*, 328, 914–921.

Molinari, F. (2002), "Missing Treatments," Evanston, IL: Department of Economics, Northwestern University.

Ord, J. (1972), *Families of Frequency Distributions*, Griffin's Statistical Monographs & Courses No. 30, New York: Hafner.

Pepper, J. (2003), "Using Experiments to Evaluate Performance Standards: What Do Welfare-to-Work Demonstrations Reveal to Welfare Reformers?" *Journal of Human Resources*, forthcoming.

Peterson, A. (1976), "Bounds for a Joint Distribution Function with Fixed Subdistribution Functions: Application to Competing Risks," *Proceedings of the National Academy of Sciences U.S.A.*, 73, 11–13.

Reiersol, O. (1945), "Confluence Analysis by Means of Instrumental Sets of Variables,"*Arkiv fur Matematik, Astronomi Och Fysik*, 32A, No.4, 1–119.

Robins, J. (1989), "The Analysis of Randomized and Non-Randomized AIDS Treatment Trials Using a New Approach to Causal Inference in Longitudinal Studies," in L. Sechrest, H. Freeman, and A. Mulley. (eds.), *Health Service Research Methodology: A Focus on AIDS*, Washington, DC: NCHSR, U.S. Public Health Service.

Robins, J., A. Rotnitzky, and L. Zhao (1994), "Estimation of Regression Coefficients when Some Regressors Are Not Always Observed," *Journal of the American Statistical Association*, 89, 846–866.

Robinson, W. (1950), "Ecological Correlation and the Behavior of Individuals," *American Sociological Review*, 15, 351–357.

Rosenbaum, P. (1995), *Observational Studies*, New York: Springer-Verlag.

Rosenbaum, P. (1999), "Choice as an Alternative to Control in Observational Studies," *Statistical Science*, 14, 259–304.

Rubin, D. (1976), "Inference and Missing Data," *Biometrika*, 63, 581–590.

Ruschendorf, L. (1981), "Sharpness of Frechet-Bounds," *Zeitschrift fur Wahrscheinlichkeitstheorie und Verwandte Gebiete*, 57, 293–302.

Savage, L. (1954), *The Foundations of Statistics*, New York: Wiley.

Scharfstein, D., A. Rotnitzky, and J. Robins (1999), "Adjusting for Nonignorable Drop-Out Using Semiparametric Nonresponse Models," *Journal of the American Statistical Association*, 94, 1096–1120.

Simpson E. (1951), "The Interpretation of Interaction in Contingency Tables," *Journal of the Royal Statistical Society B*, 13, 238–241.

Stafford, F. (1985), "Income-Maintenance Policy and Work Effort: Learning from Experiments and Labor-Market Studies," in J. Hausman and D. Wise (eds.), *Social Experimentation*, Chicago: University of Chicago Press.

U.S. Bureau of the Census (1991), "Money Income of Households, Families, and Persons in the United States: 1988 and 1989," in *Current Population Reports*, Series P-60, No. 172. Washington, DC: U.S. Government Printing Office.

Wald, A. (1950), *Statistical Decision Functions*, New York: Wiley.

Wang, C., S. Wang, L. Zhao, and S. Ou (1997), "Weighted Semiparametric Estimation in Regression Analysis with Missing Covariate Data," *Journal of the American Statistical Association*, 92, 512–525.

Wright, S. (1928), Appendix B to Wright, P. *The Tariff on Animal and Vegetable Oils*, New York: McMillan.

Zaffalon, M. (2002), "Exact Credal Treatment of Missing Data," *Journal of Statistical Planning and Inference*, 105, 105–122.

Zidek, J. (1984), "Maximal Simpson-Disaggregations of 2 × 2 Tables," *Biometrika*, 71, 187–190.

Index

Springer Series in Statistics   *(continued from p. ii)*